CHANGING PERSPECTIVES

CHANGING PERSPECTIVES

Mauricio Castillo

Cover illustration by Merrick E. Castillo

Copyright © 2015 Mauricio Castillo
All rights reserved.

ISBN: 1516971094
ISBN 13: 9781516971091

Table of Contents

Foreword

For 8 years, from 2007-2015, the American Journal of Neuroradiology was fortunate to have as its Editor-in-Chief, Mauricio Castillo, a man of unending energy and curiosity, whose intellect and unbounded interest in the world around us allowed him to combine his knowledge of medicine, history and neuroimaging to forge some of the best editorials ever to grace the pages of a scientific journal.

To produce, month after month, elegant and meaningful editorials is no mean feat, but he did just that. As one opened each issue of the journal and read those monthly editorials, constant thoughts arose, "what was the inspiration to write on this subject?, what was the spark to make him delve deeply into the subject matter?, and how could he so deftly tie all this into the neurosciences?" I asked myself those questions many times over the past 8 years, and I still remain astounded by Mauricio's probing inquisitiveness and the elegant wording which he delivered issue after issue.

Encouraged to publish an anthology of his editorials, Mauricio has selected 46 of his writings for us to enjoy once again. Clearly and brilliantly written, often with humor, the topics touch, in one way or another, the lives of all of us.

As you will see, these editorials are divided into 8 categories: Sensing and Behaving; Intelligence and Memory; Personal Improvement; Famous People; Decisions; Technology; Science of Brain Imaging; Words and Writing. For those who didn't read any or most of Mauricio's editorials, here is your opportunity to dive into some of the most enlightening, thought-provoking, and educational pieces you can find. For those who have read all of these, here is an opportunity to re-read them. For everyone, this may entice you to go back to the AJNR website and read the 37 editorials not included in this book.

The writings in the anthology are born out of sheer brilliance. One would expect nothing less from Mauricio Castillo.

Robert M. Quencer, MD

Preface

This short book contains my favorite "Perspectives." These short essays where published monthly during the time that I was editor in chief of the American Journal of Neuroradiology a "Monthly" which specializes in imaging of the nervous system. I wanted to bring a warmer more human angle to the highly specialized contents of the journal, so every month for 8 years I wrote and published in it a short essay. This book does not contain all of them but only the ones which I thought would fit a book format better. Rather than reprinting them chronologically I have loosely group them in general categories. Unlike other authors, I have decided not to revise them but to leave them as they were originally published. Readers who are not physicians and those who are not radiologists or work within the neurosciences will find them of interest. These writings attempt to bridge the science and art of neuroimaging with our everyday life and address not only scientific questions but also general and philosophical ones. The last section summarizes some of my thoughts regarding the world of editing, writing, preserving knowledge and the overall complexity of communicating in the globalized world.

I am an avid reader of the New Yorker magazine and such took a similar approach to my Perspectives. That is, all facts were checked and re-checked to assure that they are correct (any oversights are entirely my fault). According to the topic, sometimes the incubation period of a particular essay was as long as six months. Most editors hate extra and superfluous words and while writing these essays I tried to keep my sentences as short as possible and because many of the journal's readers are not native English speakers, I tried to convey my thoughts as simple as possible.

Neuroimaging is just more than what its name implies, it is a specialty that tries to answer an old question: how does the brain work and what happens to it when something goes wrong. Understanding the brain and its function is, to me, akin to understanding the universe and our nature and meaning as humans and these questions are perhaps unanswerable by science alone. As our brain tries to understand the world that surrounds it, this world changes it. A fact or a thought seems to change according to the perspective under which it is employed and examined thus the title of this collection: Changing Perspectives.

Mauricio Castillo
Chapel Hill, North Carolina September 2015

Sensing and behaving

COLD AND HOT

Most healthy human beings maintain a core temperature of 37.0°C (98.6°F) regardless of surrounding environmental conditions. Of course, our temperature normally varies slightly throughout the day and night, and this depends on when, how, and where one measures it. Normal human temperature fluctuations tend to be of about 0.5°C up or down. A healthy human being is said to be "normothermic" or "euthermic."[1] Our temperature fluctuates with the circadian cycle as in all other living organisms (circadian rhythms occur in 24-hour cycles [circa _ around and dian: day]), and in this way, our bodies can keep track of the duration of normal days over our life span.[2]

The circadian clock is regulated mostly by daylight, and when in dark and/or cold environments, organisms compensate by altering their temperatures. Retinal stimulation by daylight is directly transmitted to the suprachiasmatic nuclei in the hypothalamus, which regulates temperature.[3] When our temperature increases, hypothalamic neurons induce sweating and cutaneous vasodilation to dissipate heat. When we get cold, the hypothalamus does the opposite and induces shivering that increases heat production. Thirst is also controlled by the hypothalamus, so if our temperature increases, we lose fluids and increase our serum sodium level. Osmotic receptors in the supraoptic nuclei not only tell our cerebral cortex that we need to drink water but stimulate the neurohypophysis to produce an antidiuretic hormone that directly acts on the renal glomeruli, increasing water reabsorption.*

Thirst control involves the fore- and hindbrain (especially the area postrema). As serum concentration of sodium increases, phylogenetically ancient brain regions (the cingulate gyrus, third ventricle, orbitofrontal regions, thalami, midbrain, and hypothalamus) become activated on fMRI.[4]

Our temperature is lowest about 2–3 hours before waking up, so it is not surprising that during the wee hours of the night we commonly reach for our covers. Conversely, our temperature is higher in the late afternoon (and this is why fever and malaise peak at this time when sick). Sleep deprivation, even short-term, lowers our body temperature, and I remember feeling cold on the days after I had been up all night as an intern and sitting in the sun to warm up. The endogenous substances that most affect our circadian cycle are hormones, and temperature variations induced by them have been used to predict ovulation during the normal menstrual cycle (also called the circamensal cycle [circa_ around, mensa: month]). Contraceptives suppress the circamensal cycle and result in elevations of temperature of about 0.5°C throughout the entire month.

As mentioned previously, our temperature is highest in the afternoons when it reaches about 37.7°C. Anything above this threshold is considered as fever. Under normal conditions, our bodies are capable of dissipating heat to just about 40.0°C, and above this threshold, one enters the state of hyperthermia and starts to develop severe headache and altered mental status. Severe and prolonged hyperthermia results in a heat stroke. Neurologic abnormalities induced by heat stroke involve the brain (especially the basal ganglia and thalami), cerebellum, anterior horn cells in the spinal cord, and peripheral nerves.[5] Hemorrhages in the external capsule and medial thalami may occur. In the cerebellum, the Purkinje cells are especially sensitive to heat, and patients with hyperthermia may later develop chronic cerebellar atrophy.[6]

The opposite extreme is hypothermia, which is defined as a drop of 2.0°C in body temperature (we begin to shiver at about 36.0°C). Hypothermia is generally due to exposure to inclement weather or is induced for medical procedures. Protective hypothermia is mainly used for the treatment of neonatal encephalopathy, cardiac arrest, neurogenic fever, and more recently brain and spinal cord trauma. For these conditions, the goal is to reach 33.0°C, with the hope that slowing cellular metabolism will be neuroprotective. Lower temperatures inhibit the influx of ions (noticeably calcium) into cells, avoid oxidative stress, decrease free radical production, and, finally, prevent apoptosis. Cooling may be achieved via invasive means (catheters that pump cold saline solution into veins) or noninvasively (water blankets and cool caps that contain crylon gel at_30°C). Cool caps are used for neonatal encephalopathy but also for less serious problems such as alopecia induced by chemotherapy (in the latter, the cold results in decreased scalp blood flow protecting the hair follicles).[7]

In an intriguing new article published in the *American Journal of Neuroradiology* (*AJNR*), the authors used catheters placed in the subarachnoid space of swine to cool down the spinal cord without cooling the entire body.[8] This and other similar experiments have shown that the technique is feasible and may spare complications induced by whole-body hypothermia.

We humans detect temperature changes due to stimulation of our peripheral thermoreceptors found in the skin and mucosa.

These receptors are nerve fibers that work in tandem—that is, when exposed to cold, delta fast-speed fibers fire more than the slower C-fibers that are sensitive to heat, but both work simultaneously and the perception of temperature depends on the proportion of each type of fiber that is activated. These fibers enter the spinal cord and ascend in the spinothalamic tracts to the posterolateral thalamic nuclei. From there, the stimuli reach a collection of organs and nuclei best known as the hypothalamus.

Time for an anecdote: Years ago the Chief of Neurosurgery and I decided to seal a CSF leak due to a previous transsphenoidal pituitary adenoma resection with liquid fibrin glue. Under CT guidance, we proceeded to place a needle through the nose of the patient into the breached sellar floor and started injecting the glue.

The patient immediately became severely hypothermic, and a control CT showed that our needle was too deep and the glue had reached the hypothalamus. The patient spent a week in the intensive care unit, and though he completely recovered, it still gives me goose bumps to think about it.

There is no consensus about the most comfortable temperature.

The way we perceive temperature has a lot to do with the humidity accompanying it. For example, if the humidity is 0%, 24°C will feel like 21°C, while with a 100% humidity, 24°C will feel like 27°C. The most comfortable humidity levels are between 40% and 50% (levels also said to prevent upper respiratory tract infections).

In places with extreme outside temperature variations, it is recommended that inside temperatures be kept at 21°–23°C (69°–73°F).[9] In the United States, the Occupational Safety and Health Administration recommends a range of 20.5°–24.5°C (68°–76°F) and humidity between 20% and 60% at workplaces.[10]

Another anecdote: Upon arriving in Panama City, my colleague and friend, Dr Ilka Guerrero asked me if I had brought a sweater because the city was about the coldest place on earth. After entering my hotel, I understood what she meant. Air conditioning thermostats were kept at 16°C (62°F). Those of us who grew up elsewhere in warmer latitudes never cease to be amazed by how cold buildings are kept in the United States (they are even colder in the United Kingdom where regulations dictate temperature at the workplace be "reasonable"—that is, 16°C (61°F)![11]

Cold inside buildings is a luxury, and it has been documented that the most expensive stores (Hermes, 20°C) are kept colder than the least expensive ones (Old Navy, 27°C).[12] Apparently colder spaces encourage us to buy more and also save electricity. We radiologists must work in cold environments because computers and monitors (not to forget the view boxes of the past) generate a lot of heat and reading rooms can only be kept comfortable by lowering the thermostat. Work productivity is better at stable and slightly lower temperatures[13] but colder temperatures are uncomfortable and the New York Public Library allows their workers to accrue compensatory time when its temperature drops below 20°C (68°F).

It is a common belief that heavier individuals are more sensitive to heat (and conversely will feel cold to the touch), while skinny ones are sensitive to cold (but will be hot to the touch). Subcutaneous fat serves as an insulator, but women, who as a general rule have more of it, are more sensitive to cold than men.[14] Malnourished individuals with little fat may experience hypothermia in temperatures of only 15°–18°C.[15] An intriguing observation is that hot and cold result in nearly identical brain responses. (When ice first arrived in Macondo in Garcia Marquez's *One Hundred Years of Solitude*, people could not tell if it was hot or cold.) With fMRI, the response to noxious hot and cold stimulation was studied, and it was shown that extreme temperatures both activated similar networks.[16] While this study and others confirm the activation of unified neural networks for different intensities of temperature, other studies show that the perception of pleasantness or unpleasantness associated with temperature changes occurs in different brain regions.[17] More or less neuronal firing in these areas occurs as temperature changes.

Our feeling of well-being is tied to what we believe is a comfortable temperature, and some of our activities are immediately associated with temperature. Reading by a source of warmth such as a fireplace comes to mind except when it comes to *AJNR*, which I think can be read when it is cold or hot.

* NB: I recommend the following wonderful review dealing with issues related to thirst: McKinley MJ, Johnson AK. *The physiologic regulation of thirst and fluid intake.* *News Physiol Sci* 2004;19:1– 6.

REFERENCES

1. Human body temperature. http://en.wikipedia.org/wiki/Human_body_temperature. Accessed February 21, 2013
2. Circadian rhythm. http://en.wikipedia.org/wiki/Circadian_rhythm. Accessed February 21, 2013
3. Hypothalamus. http://www.neuroanatomy.wisc.edu/coursebook/neuro2(2).pdf
4. Denton D, Shade R, Zamarippa F, et al. **Neuroimaging of genesis and satiation of thirst and the interceptor-driven theory of origins of primary consciousness.** *Proc Natl Acad Sci U S A* 1999;96:5304–09
5. McLaughlin CT, Kane AG, Auber AE. **MR imaging of heat stroke: external capsule and thalamic T1 shortening and cerebellar injury.** *AJNR Am J Neuroradiol* 2003;24:1372–75
6. Albukrek D, Bakon M, Moran DS, et al. **Heat stroke-induced cerebellar atrophy: clinical course—CT and MR findings.** *Neuroradiology* 1997;39:195–97

7. Hypothermia cap. http://en.wikipedia.org/wiki/Hypothermia_cap. Accessed February 21, 2013
8. Purdy PD, Navakovic RL, Giles BP, et al. **Spinal cord hypothermia without systemic hypothermia.** *AJNR Am J Neuroradiol* 2013; 34:252–56
9. Canadian Centre for Occupational Health and Safety. Thermal comfort for office work. http://www.ccohs.ca/oshanswers/phys_agents/ thermal_comfort.html. Accessed February 21, 2013
10. United States Department of Labor. Office temperature/humidity. http://www.osha.gov/pls/oshaweb/owadisp.show_document?p_ table_ INTERPRETATIONS&p_id_24602. Accessed February 21, 2013
11. Buettner R. **In New York City, a chilly library has its rewards.** *The New York Times.* http://www.nytimes.com/2010/01/12/nyregion/12libraries. html?_r_0. Accessed March 11, 2013
12. Salkin A. **Shivering for luxury.** *The New York Times.* June 26, 2005. http://www.nytimes.com/2005/06/26/fashion/sundaystyles/26air. html. Accessed March 11, 2013
13. Salmon R. **Air-conditioning made this debate possible.** *The New York Times.* http://www.nytimes.com/roomfordebate/2012/06/21/ should-air-conditioning-go-global-or-be-rationed-away/airconditioning- made-this-debate-possible. Accessed March 11, 2013
14. Fillingim RB, Maixner W, Kincaid S, et al. **Sex differences in temporal summation but not sensory-discriminative processing of thermal pain.** *Pain* 1998;75:121–27
15. Human biological adaptability: an overview. http://anthro.palomar. edu/adapt/adapt_1.htm. Accessed February 21, 2013
16. Tracey I, Becerra L, Chang I, et al. **Noxious hot and cold stimulation produce common patterns of brain activation in humans: a functional magnetic resonance imaging study.** *Neurosci Lett* 2000; 288:159–62
17. Rolls E, Graberhorst F, Parris BA. **Warm pleasant feelings in the brain.** *Neuroimage* 2008;41:1504–13

THE COMPLICATED EQUATION OF SMELL, FLAVOR, AND TASTE

Our oldest senses are those related to chemogustatory capacities: smell and taste. Of these, smell is probably the oldest, and before we fully develop cerebral hemispheres, the olfactory apparatus already exists as extensions of the limbic system. The study of the senses of smell and taste is so complex that it encompasses armies of aromachologists, food scientists, physiologists, behavioral psychologists, cognitive neuroscientists, neuropharmacologists, biochemists, anthropologists, molecular biologists, and many more and is intimately related to the study of taste.

The olfactory system in vertebrates has a unique embryology.

It forms from 1) paired placodes made of non-neural epithelium that have the capacity to give rise to sensory neurons and supporting cells in the olfactory epithelium, and 2) neural crest cells that give origin to the structural elements of the nose and its cavities.[1] Although one cannot form without the other, neural crest cells get to their destination first. The olfactory receptor neurons are in the nasal cavity, and their axons, arranged in fascicles, traverse the cribriform plates and dura to synapse with cells in the olfactory bulbs, which are extensions of the brain. The olfactory neurons and accompanying glial cells arise outside the central nervous system but have the capacity to regenerate throughout life; it seems that progenitor neural crest cells may be their origin.

The human sense of smell is bidirectional, and the way we perceive smells changes according to the direction of air flow.

Orthonasal smell is perceived when breathing in, while retronasal smell occurs when odorized air arising from the mouth is forced into the nose. This last type of smell is much more complex than the first one because it recruits flavor, texture, hearing, and muscle activity. Animals with a great sense of smell like dogs are designed predominantly for orthonasal smell. Their long snouts concentrate, moisturize, and direct odorized air directly toward their olfactory epithelium, assuring that warmed molecules are easily detected. Dogs have over 220 million olfactory receptors (compared with 5–10 million in humans), a 40% greater area of the brain dedicated to smell, and the ability to smell 1000–10,000 times better than humans.[2] In addition, dogs have a large vomeronasal (Jacobson) organ, whose neurons extend to accessory olfactory bulbs and then to the hippocampi. This organ is predominantly involved in pheromone perception and does not play an important role in the human sense of smell. Pheromones have a "blind smell," meaning that they stimulate the brain (observed with fMRI) while having no odor that can be perceived. Females are sensitive to male pheromones, particularly during ovulation.

Male pheromones are found in sweat, but only fresh sweat. After 20 minutes, sweat is oxidized and it just smells bad. Billing agencies will send out bills scented with androstenone (a pheromone) because they are then perceived as being more aggressive and increase their collection rates[3] (note: the report in this reference is wonderfully entertaining!).

Because humans mostly depend on vision that is stereoscopic, which in turn is contingent on a strict interocular distance, we do not have a long snout and our anterior nasal pathways are less complex, less efficient, and shorter than those of dogs. However, contrary to prior beliefs, there is no evolutionary competition between smell and taste and vision; our sight has improved but our sense of smell remains quite good and the blind do not have a better sense of smell than the sighted. Although most mammals depend on orthonasal smell, we humans mostly use retronasal smell. When we say something tastes good, in reality we mean that it smells good because most "flavor" is actually retronasal smell. While retronasal smell is essential for tasting, antegrade smelling is not.

Once odorized air enters the nose in antegrade or retrograde directions, it reaches the cilia of the olfactory neurons where about 1000 specific receptor proteins are present. Specific olfactory receptor genes encode each protein. The discoverers of these genes were awarded the 2004 Nobel Prize.[4] Because humans can see very well, we do not depend on smell too much. The development of tricolor vision led to many olfactory receptor genes disappearing.

[5] Humans can still differentiate about 10,000 smells, but to name them all, you have to be an expert such as wine connoisseur Robert Parker. The molecules for each smell have individual chemical and physical configurations that allow them to bind with specific receptors (the so-called "lock and key" concept).

Once the molecules bind, adenylate cyclase is stimulated, and the result is an electrical impulse carried to the mitral cells that reside in the olfactory bulbs and send axons to different parts of the brain as follows:

- The piriform cortex is formed by the amygdala, uncus, and parahippocampal gyrus and is involved in perception of smells.
- The entorhinal cortex is the anterior aspect of the parahippocampal gyrus. Its function is to pair specific odors to specific memories (remember that an abnormal sense of smell is typical of Alzheimer and Parkinson disease).[6,7]
- The olfactory tubercle is located close to the nucleus accumbens; it is not directly involved in the perception of smells but rather in reward behaviors associated with odors.
- The amygdala is involved in emotional and autonomic responses to odors.

A good sense of smell can make up for the loss of taste as seen in the case of Chef Grant Achatz. Mr Achatz's Chicago restaurant, Alinea, now ranks as the sixth best in the world (it also has been awarded 3 Michelin Stars). Mr Achatz developed and neglected an oral cancer until it became stage 4, and rather than losing his tongue, he decided to undergo chemotherapy, radiation, and surgery.[8] Although his cancer is in remission, he lost his sense of taste (but not his sense of smell). I have eaten at Alinea and can assure our readers that this handicap has not affected the taste of the wonderful food he there designs and prepares. The relationship between smell, taste, and the brain is studied by a discipline called "neurogastronomy."

Smells produce activation in specific olfactory bulb regions, depending on their specific chemical compounds and timing. It seems that olfactory bulbs use mechanisms similar to our "visual pattern recognition" to identify discrete smells. "Odor images" refer to maps of olfactory bulb activity during olfaction. The SenseLab Web site contains many of these maps obtained with fMRI at 7T in animals (because the olfactory bulb contains neurons, their activation may be mapped by fMRI as we commonly do in the brain).[9] Curiously, smell perception may occur after all related brain areas are damaged, as long as the olfactory bulbs remain intact.

Similar to the olfactory system, taste depends on the specific recognition of different flavors by specific cells in the tongue.

We Westerners recognize 4 stimuli (salt, acid, sweet, and bitter), while Asians can add Umami (savory or meat-like) to these. Unlike the olfactory cells, taste does not get to the brain directly but via the seventh, ninth, and tenth cranial nerves. It is unclear whether stimuli traveling through these nerves compete and complement each other, but the end of the road is the insular cortex, where taste becomes a conscious activity. Weak smells and taste are congruent with each other, and they add up to recognizable flavors. Molecular cuisine, like that practiced in Alinea, El Bulli, and other famous modernistic restaurants, is based on the combination of non-congruent ingredients that result in a new perception of flavors. For example, Adam Melonas, a disciple of El Bulli's Chef Ferran Adria, created the "Octopop," a lollipop made of sliced, orange-infused octopus.[10] In this dish, the strong retronasal smell of oranges mixed with the texture of the octopus leads to sensory fusion and a greater recruitment of brain regions needed to analyze what is in one's mouth, resulting in a totally new experience.

Chefs can only go so far because most tastes and smells have strong emotional (hedonic) components that may render some combinations repellent. Last year, my wife and I decided to eat at Corton's in New York and found ourselves disliking their utterly strange combination of flavors. Sensory fusion overload did us in, but what we disliked may have been pleasurable to others, particularly individuals (supertasters) who may be able to taste the individual components in a dish and enjoy them for what they are. When aiming for sensory fusion, one needs exact amounts of ingredients (that is why molecular cuisine is considered very close to chemistry). For example, add too much capsaicin and this irritant suppresses the taste of all else in a bite.

Although the title of this *Perspectives* refers to the equation "smell _ taste _ flavor," one must also add "mouth sense" to make it correct. Mouth sense is important to taste and refers to temperature, pain, touch, and pressure receptors inside the oral cavity. These sensations are transmitted to the somatosensory cortex via the trigeminal nerves. The cornerstones of molecular (also called modernistic) cuisine—foams, spheres, and powders—rely on producing a flavor with a totally different mouth sense to surprise us. Next, we must add vision to the equation. Vividly colored food tastes and smells more intense than bland colored food.a Conditioning also plays a role, and if, for example, white wines are colored red, many tasters will believe that they were given red wines. The last addition to the equation is hearing. If we buy a cracker said to be "crispy," it must produce a sound of more than 5 kHz in our mouth for us to perceive it as such. Carrots, which we expect to be "crunchy," produce a sound between 1 and 2 kHz.

In his book *Neurogastronomy,*[11] Gordon M. Shepherd, a professor of neurobiology at Yale University, refers to the "flavor perception system."b Professor Shepherd calls the other half of the equation (emotion _ memory _ decisions _ plasticity _ language _ consciousness _ flavor) the "flavor action system," and one system will not work without the other. Scientists who study addictions (particularly drug and food ones), as well as neuroeconomists, are very interested in these systems. On the basis of understanding both, the food industry has created the most addictive and universal food item: the French fry (buttery smell, salty taste, crunchy feel and sound, and a vivid yellow color).

The industry also knows that children prefer sweet and salty over sour and bitter. Colors do strange things to flavor and our desire to eat. Orange and yellow are said to induce appetite and thus are used in the decor of many fast food restaurants. Green, brown, and red are the most used colors in the food industry because they are the naturally occurring ones and we associate them with nature and thus, health. Blue is linked to sweetness, but it is not a natural food color and we tend to avoid food of this color. Colors and taste have no relationship with nutritional value. Smells cause strange sensations; who would have imagined that the smell of pumpkin pie increases penile blood flow?

As we age; our senses of smell and taste deteriorate, a finding nearly universal after 60 years of age, and we are all familiar with the consequences of adding too much salt or sugar to our food. By 80 years of age, most individuals' sense of smell is significantly impaired (though women fare much better than men). Cuisine for the elderly attempts to compensate for decreasing senses by making food more palatable. Regardless of age, the flavor equation is complex, and I like to think of it as follows:

> (smell + taste + mouth sense + sight + sound) X (emotion + memory + decisions + plasticity + language + consciousness)= flavor.

a "Synesthesia" refers to an ability to see color or hear sounds when smelling particular odors. Some individuals are gifted and have a broad sense of synesthesia (such as perfume makers); however, all of us have some synesthetic ability.

b Although this book can be found at your local Barnes and Noble, beware: It is an academic and serious work that requires some serious effort to read and understand.

REFERENCES

1. Katoh H, Shibata S, Fukuda K, et al. **The dual origin of the peripheral olfactory system: placode and neural crest.** *Molecular Brain* 2011;4:34–50
2. The canine sense of smell. http://www.whole-dog-journal.com/issues/7_11/features/Canine-Sense-of-Smell_15668–1.html. Accessed August 14, 2013
3. The smell report. http://www.sirc.org/publik/smell.pdf. Accessed August 14, 2013
4. Buck L, Axel R. **A novel multigene family may encode odorant receptors: a molecular basis for odor recognition.** *Cell* 1991;65:175–87

5. Gilad Y, Przeworski M, Lancet D, et al. **Loss of olfactory receptor genes coincides with the acquisition of full trichromatic vision in primates**. *PLoS Biol* 2004;2:E5
6. Wesson DW, Levy E, Nixon RA, et al. **Olfactory dysfunction correlates with amyloid-beta burden in an Alzheimer's disease mouse model.** *J Neurosci* 2010;30:505–14
7. Haehner A, Hummel T, Hummel C, et al. **Olfactory loss may be a first sign of idiopathic Parkinson's disease.** *Mov Disord* 2007;22:839–42
8. Grant Achatz. Wikipedia. http://en.wikipedia.org/wiki/Grant_Achatz. Accessed August 14, 2013
9. Olfactory bulb odor map database. SenseLab. http://senselab.med.yale.edu/OdorMapDB. Accessed August 14, 2013
10. MADRID LAB. A progressive cuisine lab in Spain. http://www. madridlab.net/melonas. Accessed August 14, 2013
11. Shepherd GM. *Neurogastronomy.* New York: Columbia University Press; 2012

SOUNDS OF SILENCE

"Do not speak unless you can improve silence."
Anonymous

Lately, I have noticed my fellows and residents listen to music while studying images and dictating reports. I remember, when younger, I used to do the same, but now I crave silence. When writing, editing, or preparing conferences, I prefer to do it in my office, door closed, telephone ignored.

When faced with a difficult situation, I wish for the muted silence that follows a snowfall. Some are fond of saying that silence is golden. The word "silence" comes from "silere" (unknown origin and meaning to be quiet or still), which then gave origin to the

Latin "silentium" (being silent), and after that the French "silence" (absence of sound).[1]

Silence, or rather the noise that makes it nearly impossible to find, is the topic of *In Pursuit of Silence: Listening for Meaning in a World of Noise* by George Prochnik.[2] Prochnik says America is getting louder and younger generations are becoming addicted to sounds and noise. Noise can serve as a stimulant, but in large doses, it is counterproductive. Noise was abundant during the Industrial Revolution but not as ubiquitous and constant as it is now. These days, background noise is everywhere, from the elevator and the mall, to our cars, homes, and offices. Noise fragments your sleep and makes you feel tired the next day; your heart beats faster, you get vasoconstriction, and though the brain manipulates it to the point where noise becomes "invisible," your body continues to feel stressed. Sometimes we are forced to choose between noises, one used to disguise another (like wearing your iPod). This is the case with "white noise," which is generally used to mask other sounds. The term "white noise" is adapted from "white light," in which the retinal cones for green, blue, and red (the primary colors) are equally stimulated, perceiving light as "white" or colorless. Thus, by extension, white noise is a sound you cannot perceive. White noise is evident in devices such as noise-cancelling headphones, which mask environmental and unwanted sounds (a case of creating noise to abate other noise). Curiously, noise comes in different "colors" such as pink (used as a reference sound), blue (low frequency without spikes), purple (a type of white noise), gray (equally loud at all frequencies), and the so-called unofficial noise colors (black, green, orange, and red).[3] Electricity has its own noise called the "Mains hum."[4]

With so many different types of noise surrounding us, it is not surprising that some individuals have become "anti-noise activists."

Anti-noise militants wage battles against dogs barking, vehicle traffic, and airplanes, among other sounds. Dog barking is particularly annoying, and the authorities know this and use it to their advantage. Remember, the Branch Davidians in Waco, Texas, and Dictator Manuel Noriega in Panama were driven out of their sanctuaries with the aid of noise bombardment by way of dogs barking.[5] If wolves, which are the ancestors of dogs, do not bark, why do dogs do it? In his book, *Dogwatching*, the famous and popular zoologist, Desmond Morris proposes this is the result of thousands of years of selective breeding by humans that have resulted in super barking animals.[6] Nomads use dogs as alarms. Noises of distress, coming from all kinds of animals and humans, are particularly stressful. A British social anthropologist, Sheila Kitzinger, says, "The sound of a crying baby…is just about the most disturbing, demanding, shattering noise we can hear."[7] Taking advantage of this, the military used nonstop baby crying recordings as torture in the prison at Guantanamo Bay. The most frequent complaint Americans have about their neighborhoods is noise, not crime.[8]

Although as physicians, we may focus on health-related problems induced by noise, the government sees the issue as one related to civil liberties. In 1972, Congress passed the Noise Control Act by which the Environmental Protection Agency was put in charge of studying ways to reduce noise. In 1982, the Office of Noise Abatement and Control was closed.

Today, there is an interest in re-establishing the Noise Control office of the Environmental Protection Agency, albeit with a very small budget.[9] The United States is one of the few developed countries in which government has no control over noise. The European Commission on the Environment has issued the European Noise Directive that charges all states to inform the public about the dangers of noise and related issues.

Ways to reduce noise have been carefully studied in Europe where cities are smaller and living quarters closer to roads. Noise-induced stress is thought to be responsible for 3% of heart attacks in Germany.[10] Overall, about 45,000 fatal heart attacks occur worldwide as a consequence of noise-induced stress. As population density increases, so does noise.

When population density reaches 10^5 per square mile, the average background noise level is about 75 dB (85 dB is considered harmful and may result in permanent hearing deficits).[11]

Reductions in noise levels as small as 3 dB can make a difference (remember that the decibel scale is logarithmic).

Biologic effects of noise may be aural and non-aural. Non-aural effects include impairments in communication, sleep, recreation, and performance along with an increased sense of annoyance that leads to cardiovascular disease. Individuals exposed to transportation noise have an increased risk of hypertension and ischemic heart disease.[12] Prisons in which noise is abundant experience higher levels of inmate aggression than quieter ones. Perception of noise is somewhat cultural. For example, the Polish report increased annoyance due to traffic noise compared with Austrians, and people living in Munich are much more annoyed by noise than those residing in Genova.[10] Noise mapping for some major cities is available. A noise map for the city of San Francisco (second in terms population density after New York) indicates the Union Square and Embarcadero zones are the noisiest.[13] The municipality of the city of Paris offers an interactive, 2D and 3D, real-time noise map that presumably helps you to avoid the noisiest parts of La Ville Lumiere. Noise maps are not only used in cities but also in smaller environments such as factories to determine their loudest parts. By definition, "noise" is a sound that is loud, unpleasant, unexpected, or undesired.[14] The origin of the word noise is probably from the Latin "nausea," which means sickness.

The auditory cortex is involved in both processing noise and silence. Recently, it has been postulated that the impulses arrive at the auditory cortex via different pathways.[15] A spe- cific pathway is turned on when a noise/sound occurs, but to turn it off, we need to stimulate a completely different set of neurons. If that does not happen, we cannot perceive the end of a sound, and our brain has no time to process and understand it. Thus, hearing problems can be due to the fact that we cannot activate the ON or OFF auditory circuits. The ability to hear someone speak to us at a loud party depends on this. A voice needs to about 15 dB louder (at a distance of 3 feet) than the background noise for our brain to keep the ON circuit working while stimulating the OFF circuit. Once the latter is activated, background noise becomes less intrusive and we can concentrate on what an individual is saying. If we cannot activate the OFF circuit, we will be unable to separate an individual's voice from the background din. For hearing and speech therapists, this is of tremendous importance: instead of devices reinforcing the activity of the ON circuit, devices that activate the OFF circuit may help patients improve hearing.

Silence is critical to comprehension and learning. In 2002, a study of 326 children (mean age, 10 years) who were exposed to airport noise was published.[16] Reading, long- and short term memory, and speech perception were all significantly affected by noise. Furthermore, with cumulative noise exposure, the children's ability to read worsened. Similar deleterious effects have been found in school children exposed to road traffic noise.[17] Those of us who have learned a new language as adults know silence is critical to understand it. Competing speech and background noise make it very difficult to understand the spoken word.[18] In those instances, we revert to lip reading.

Throughout history, silence has been associated with higher spiritual and intellectual states. Creativity needs silence, and when James Joyce faced writer's block, he wrote to his brother, "No pen, no ink, no table, no room, no time, *no quiet* [italics mine], no inclination."[19] Times of reflection call for silence. Cerebral blood flow increases and neural networks are more efficiently activated during times of silent meditation.

[20,21] When compared with the East, we Westerners feel somewhat uncomfortable with silence. This is particularly true in the United States, where even the Apaches and other Native American tribes used silence as means of expressing anger and other negative feelings. In spirituality, silence does not refer to an outside state but rather one of inner peace. Silence is revered in the Christian, Buddhist, Hindu, and Islamic faiths.

The gods speak to us by using silence rather than earthly words. Religious individuals are known to take a "vow of silence," and in some monasteries, the night hours are called "great silence" because speaking is forbidden. Not surprisingly, the concept of silence is closely associated with death.

Do we better understand silence as we become older because death is closer? Death leads to the ultimate state of inescapable silence. To honor the dead, we generally have a moment of silence. A minute of silence is held during Remembrance Day (November 11 at 11:00 AM) to honor those killed in wars.

Every mid-April, a Day of Silence is held to protest against injustices toward gays and lesbians. Silence may be used to mask unspeakable acts (e.g., conspiracy of silence, code of silence).

In music, a period of silence is called a "rest" and is considered essential to distinguish between different parts of a composition.

Silence in music is used to raise listeners' expectations about what comes next. Of course, for the sake of shocking audiences, this has been taken to extremes. In 1952, minimalist composer John Cage premiered his 3-act score called *4_33*.[22] This composition, written for any instrument, instructs musicians not to play, and thus the only sounds heard are those arising from the surrounding environment. Of silence, Cage said, "There is no such thing as an empty space, or an empty time, try as we may to make a silence, we cannot."

Years later, John Lennon and Yoko Ono wrote a composition that sounds exactly like Cage's and is called "Two Minutes of Silence." The word silence has also been used in innumerable popular songs (e.g., "Silent Night," "The Sounds of Silence"

[Simon and Garfunkel], "The Silence" [Depeche Mode]).

Complete silence cannot be achieved unless you are deaf.

Right now I am typing in a relative silence; the air conditioning hums, the keyboard clicks, I can hear my own breathing. Silence is an essential part of life that I fear is disappearing. As I was lying in bed last night, the idea of silence was so strong that I decided to write this short essay on it. To finish, I quote *The Sounds of Silence*:

"Hello darkness, my old friend I've come to talk with you again Because a vision softly creeping Left its seeds while I was sleeping And the vision that was planted in my brain Still remains Within the sound of silence"

References

1. Dictionary.com. www.dictionary.reference.com. Accessed July 4, 2010
2. Prochnik G. *In Pursuit of Silence: Listening for Meaning in a World of Noise*. New York: Doubleday; 2010
3. Wikipedia. **Colors of noise.** http://en.wikipedia.org/wiki/Colors_of_noise. Accessed July 7, 2010

4. Wikipedia. **Mains.hum.** http://en.wikipedia.org/wiki/Mains_hum. Accessed July 7, 2010
5. Science Dog Network. **BarkingDogs.net.** www.barkingdogs.net. Accessed June 3, 2010
6. Morris, D. *Dogwatching.* New York: Crown Publishers; 1987
7. Jerome Goopman. **The colic conundrum.** *The New Yorker.* http://www.newyorker.com/reporting/2007/09/17/070917fa_fact_groopman#ixzz0pnMix8TM. Accessed June 16, 2010
8. The Right to Quiet Society. **Welcome to peace and quiet.** http://www.quiet.org. Accessed June 3, 2010
9. The Free Library by Farlex. www.thefreelibrary.com. Accessed July 7, 2010
10. www.silence-.ip.org/site. Accessed July 5, 2010
11. http://www.hmmh.com/cmsdocuments/PopulationDistribution_US_Function OutdoorNoiseLevel.pdf. Accessed July 7, 2010
12. Babisch W. **Transportation noise and cardiovascular risk: updated review and synthesis of epidemiological studies indicate the evidence has increased.** *Noise Health* 2006;8:1–29
13. San Francisco City-Wide Noise Map. http://www.sfdph.org/dph/files/EHSdocs/ ehsPublsdocs/Noise/noisemap2.pdf. Accessed June 3, 2010
14. The Free Library by Farlex. www.thefreedictionary.com/noise. Accessed June 25, 2010
15. Scholl B, Gao X, Wehr M. **Nonoverlapping sets of synapses drive on responses and off responses in auditory cortex.** *Neuron* 2010; 65:412–21
16. Hygge S, Evans GW, Bullinger M. **A prospective study of some effects of aircraft nose on cognitive performance in schoolchildren.** *Psychol Sci* 2002;13: 369–474
17. Hygge S, Boman E, Enmaker I. **The effects of road traffic noise and meaningful irrelevant speech on different memory systems.** *Scand J Psychol* 2003;44:13–21

18. Hygge S, Ronnberg J, Larsby B. **Normal-hearing and hearing-impaired subjects' ability to just follow conversation in competing speech, reversed speech, and noise background.** *J Speech Hear Res* 1992;35:208–15 **1156** Editorials _ AJNR 32 _ Aug 2011 _ www.ajnr.org
19. Manguel A. *A Reader on Reading*. New Haven, Connecticut: Yale University Press; 2010:20
20. Lazar SW, Bush G, Gollub RL, et al. **Functional brain mapping of relaxation response and meditation.** *Neuroreport* 2000;11:1581–85
21. Short EB, Kose S, Mu Q, et al. **Regional brain activation during meditation shows time and practice effects: an exploratory fMRI study.** *Evid Based Complement Alternat Med* 2007;7:121–27
22. Wikipedia. **John Cage.** http://en.wikipedia.org/wiki/John_Cage. Accessed July 7, 2010

LISTENING TO MUSIC

Charlotte Gainsbourg, an actress and singer, daughter of that famous late 1960s couple (Serge Gainsbourg and Jane Birkin) has a new music CD out. It is called *IRM* (MR imaging in French) as is the hit song it contains. Ms Gainsbourg had a waterskiing accident that led to her having to undergo several brain MR imaging studies, an experience that apparently inspired her (and her producer, Beck) not only to title her album as such but also to use gradient-like sounds as a part of the rhythm lines in many of her songs (particularly in *IRM*). Having been inside the magnet several times, I wondered how this could possibly work well, but now that I have listened to the CD, I like it. The contents of the booklet that comes with it are also somewhat surprising: It contains several MR angiography views of the circle of Willis (presumably Gainsbourg's) as well as patient identifiers from the alphanumeric information of the study that include her name, date of birth, medical record number, etc, clearly a Health Insurance Portability and

Accountability Act violation. Gainsbourg and Beck are not the first modern musicians to borrow from MR imaging; other rock groups are called MRI, Tesla, The Magnetic Fields, and the more intriguing MRI Resident Research Orchestra and Rock Band.

The nature of music is universal across all cultures, but the fact that I liked her CD is a matter of perception. Functional MR imaging (fMRI) has been used to study music perception in humans.[1] Music perception involves the processing of several activities that mainly include the following: 1) musical syntax, 2) musical meaning, 3) auditory working memory, and 4) emotional aspects.

Music plays an important role in brain maturation as it relates to communication and social, cognitive, and emotional development. Similar brain regions are used for the processing of music and language.[2] So, it makes sense to start musical education at the same time we start learning a language. Music processing in the brain takes place mainly in the inferior frontal gyrus, orbital frontolateral cortex, anterior insula, ventrolateral premotor cortex, anterior and posterior superior temporal gyri, superior temporal sulcus, and supramarginal gyrus.

Activation by fMRI is seen in both hemispheres but differs with the subject's age and degree of musical training. Children show right-sided activations similar to those seen in adults, but on the left side, they show a lesser degree than adults, particularly in the prefrontal and temporal areas and in the supramarginal gyrus. More musically trained individuals show stronger activations in the frontal operculum and superior temporal gyrus, regardless of age. That both trained and untrained individuals show comparable brain activations may help explain the wide popularity of music: Our brains are wired for music, or, in other words, our brains are very interested in music. Not only is our cerebrum interested in music, but our cerebellum is also interested in it. Patients with cerebellar degeneration have impaired fine discrimination of pitch, whereas healthy individuals use their cerebellum mainly when processing rhythm (including pattern, tempo, meter, and duration).[3]

It is interesting that one does not have to listen to music to activate many of the aforementioned areas. Simply thinking about it suffices (remember, deaf musicians such as Beethoven and Smetana wrote great works simply by imagery).[4] Additionally, you can only tune an instrument if you can imagine how it should sound. Motor imagery is directly linked to auditory imagery: For example, pianists practice by imagining their hands on the keyboard. Imagery also plays an important role when music is spontaneously improvised.

Improvisation is an important part of music (for example, in jazz). In 1 study, 11 professional pianists were asked to either improvise or reproduce music by using a small piano keyboard specifically designed to fit inside a 1.5T unit.[5] During both activities, similar activations were found, but there was a trend for a larger motor output during reproduction of music and for activation of the right dorsolateral prefrontal cortex and the premotor and auditory areas. The importance of this study is that it shows that fMRI may be used to investigate the brain during the act of creation, probably one of the most important activities defining humans.

Because classically trained pianists are not prone to improvisation, what happens with jazz pianists who, according the nature of that genre, need to constantly improvise? Jazz musicians improvise extemporaneously during their solos, and that is why no 2 jazz performances are ever identical. In a different study, 6 highly trained jazz musicians were asked to spontaneously improvise by using a non-ferromagnetic keyboard while inside a 3T unit.[6] Data obtained showed that during improvisations the lateral prefrontal cortex deactivated while there was focal activation of the medial prefrontal cortex. It appears that music creation can happen outside of conscious awareness and beyond volitional control. When people say a musician appeared to be in a "trance" while playing, they are not far from the truth. Deactivation of the lateral prefrontal cortex also occurs in altered states of consciousness such as hypnosis, meditation, and daydreaming. Also during intense musical pleasure, the limbic system deactivates (inhibitions and self-censoring are turned off).

"Pleasant" and "unpleasant" music elicits different emotional reactions. In another fMRI study, the consequences of listening to permanently dissonant (unpleasant) and consonant (pleasant) music were studied.[7] A surprising result related to the temporal dynamics linked to pleasant music. In all areas of the brain involved in listening (except for the hippocampi), increased activations occurred with time. Confirming that the amygdalae are activated during instances of negative emotional valences, unpleasant music activated the anterior and mesial temporal lobes, while pleasant music resulted in decreased activations or deactivations in the same areas. Music has the capacity to up- or down-regulate neuronal activity in identical areas of the brain. These studies are of enormous interest to the music industry in the prerelease testing of new music. Fortunately, it is not that simple to find out which songs will appeal to most people because what is considered musically pleasant or unpleasant varies significantly among cultures.

The social context in which music is listened to or played also has a role in brain activation. It seems that when musicians sing together, they activate different and more parts of their brains than when they do it alone. Jarvis Cocker, a British pop star and front man for the group Pulp, had fMRI studies after singing alone or in a duet. His brain was reported to be more "active" after the duet, a finding probably related to the coordination needed and the emotional involvement that singing with another entails.[8] The ultimate consequence of all these electric and chemical activities elicited by music and particularly by playing an instrument is that of morphologic changes in the brain.

Learning-induced cortical plasticity results in changes in the primary motor and somatosensory areas: That is, repeated tasks lead to expansion of cortical representations. Professional violinists show enlargement of the left-hand representation in their sensorimotor cortex.[9] This expanded representation is more prominent in those individuals who learned to play the violin when very young. Indeed, brain reconfiguration can occur very quickly. Structural changes in the brain have been documented after only 15 months of musical training during early childhood.[10] Trained pianists show differences in the somatotopic hand area on the right central sulcus compared with controls, and these differences are marked when training began early.[11]

Of course in the case of Ms. Gainsbourg, identifying the morphologic aberration (which apparently was a hematoma) was much easier than perhaps attempting to identify what makes her a good singer and an excellent actress. "Can you see a memory, register all my fear, on a flowchart disappear, leave my head demagnetized, tell me here the trauma lies, in the scan of pathogen, the shadow of my sin?" go the lyrics of *IRM*. To many patients undergoing an MR imaging study, it may seem a haunting experience, and it is unclear if Ms. Gainsbourg found the MR imaging noise unpleasant or pleasant, but from her album, it seems that she did find it musical.

References

1. Koelsch S, Fritz T, Schulze K, et al. **Adults and children processing music: an fMRI study.** *Neuroimaging* 2005;25:1068–76
2. Koelsch S, Gunter TC, v Cramon DY, et al. **Bach speaks: a cortical "language network" serves the processing of music.** *Neuroimaging* 2002;17:956–66
3. Parsons LM. **Exploring the functional neuroanatomy of music performance, perception, and comprehension.** *Ann N Y Acad Sci* 2001;930:211–31
4. Zatorre RJ, Halpern AR. **Mental concepts: music imagery and auditory cortex.** *Neuron* 2005;47:9–12
5. Bengtsson SL, Csikszentmihalyi M, Ullen F. **Cortical regions involved in the generation of musical structures during improvisation in pianists.** *J Cog Neurosci* 2007;19:830–42
6. Limb CJ, Braun AR. **Neural substrates of spontaneous musical performance: an fMRI study of jazz improvisation.** *PLoS One* 2008;3:e1679
7. Koelsch S, Fritz T, V Cramon DY, et al. **Investigating emotion with music: an fMRI study.** *Hum Brain Mapp* 2006;27:239–50
8. Jarvis Cocker Enters MRI, Duets With Richard Hawley . . . for Science! Pitchfork. http://pitchfork.com/news/35748-jarvis-cocker-enters-mri-duetswith- richard-hawleyfor-science. Accessed January 27, 2010
9. Schwenkreis P, El Tom S, Ragert P, et al. **Assessment of sensorimotor cortical representation and motor skills in violin players.** *Eur J Neurosci* 2007;26:3291–302
10. Hyde KL, Lerch J, Noroton A, et al. **The effects of musical training on structural brain development.** *Ann N Y Acad Sci* 2009;1169:182–86
11. Li S, Han Y, Want D, et al. Mapping surface variability of the central sulcus in musicians. Cereb Cortex 2010;20:25–33

SMILES

Have you noticed how lately we have become a nation of stunning "white" smiles? Although we Americans have always been critical of the teeth of others, such as those of the French and British, at no other time in history has this difference been as evident as now. From Hollywood to the daily news to our colleagues and children, white teeth are everywhere.

Okay, I am guilty, too, and I accept it. Twice a year I go to the drugstore and buy my whitening strips, and when I am done with them, my wife makes fun of my "ghost" teeth. In my case, the search for perfect teeth started when I was very young because my mother admired beautiful American teeth (as I imagine other mothers in other Third World countries also did). So after years of orthodontics akin to medieval tortures, I feel the need to keep them as white as possible. History repeats itself, and now I do the same to my children.

Why not have beautiful white teeth when a smile can mean so much? Nice teeth increase your confidence and you smile more. Babies are said to recognize smiles within the first 6 weeks of life. At first, their smiles are nonselective and babies smile at everyone. Visual recognition happens first, but shortly thereafter, smiles become linked to auditory stimuli (congenitally blind babies smile when they hear specific sounds but their ability to smile develops later). Smiles are integral parts of our faces, and babies immediately associate both—face plus smile—and are said to recognize the smile of a mother or father at about 2 months of life.

Specific parts of our faces, such as our eyes, trigger smiles. If you look someone directly in the eyes, one of their first spontaneous reactions is to smile. If someone looks directly at your mouth, you also tend to respond with a smile. From 3 months of age on, smiles become tools of social interaction. Parents and children exchange smiles in response to emotional, auditory, physical, and even olfactory and tactile stimuli. Children often smile as they are falling asleep or during sleep. By 4 months of age, smiles become organized and predictable. At about 5 months of age, the "cognitive" smile develops, presumably an indication of self-recognition.

As such, children smile when they perceive they have done well or accomplished something. Last to come is the "mastery" smile—that is, one elicited when children feel that they have made something happen. At about 24 months of age, children are very familiar with humor and smile at jokes, but they also smile when they feel satisfaction at having breached social rules.

There are 4 different types of smiles in adults. The simple smile is just an upward curving of the mouth (as in the "happy face" illustrated in the title of this *Perspectives*). The other 3 types are variations of the same but of differing intensities. The Duchene smile involves opening the mouth and raising the cheeks. The Duplay smile includes crinkling of the eyes, and in the last type of smile—the play smile—the mandible drops and after that we cross to laughter.

Because examining the effects of smiles in infant brains is difficult, the opposite has been done: examining the effects of baby smiles on their mother's brains. In 1 study, 28 mothers were shown pictures of their babies and of other babies while happy (smiling), sad, or neutral; then they were assessed with fMRI.[1] Dopamine-associated reward-processing brain regions were activated when mothers viewed their babies, regardless of facial expression. These areas include the ventral tegmentum, substantia nigra, striatum, and frontal lobes.

When the same mothers viewed pictures of their own smiling babies, the substantia nigra and dorsal putamen were activated, something that did not occur with sad or neutral expressions.

Thus, it seems that specific brain areas are responsible for the mother-infant attachment, and even more specific areas function when a child smiles at his or her mother. Because smiles generally signify happiness and self-fulfillment, it seems logical to assume that the brain responds differently to anger or happiness.

In another study, PET showed that facial expressions associated with anger elicited predominantly right-sided responses involving the medial, superior, middle, and inferior frontal lobes and cerebellum, whereas smiles increased activity in the cuneus, temporal lobe, and middle, medial, and superior frontal lobes.[2] Dynamic and static expressions of happiness also result in activity in different brain regions—that is, seeing someone smile in the movies (or in real life) or in a static photograph results in different brain activations. Since smiles may be fleeting, capturing one at its best in a photograph is not trivial.

Because we want to be photographed with our best smile, newer cameras have smile-recognition software that allows us to capture that fleeting moment when teeth flash. This ability started with facial-recognition programs that later were tailored to identify smiles. Once a camera detects a smile, the shutter is activated. This capability makes a camera somewhat more expensive but allows you to capture spontaneous smiles.

Computers recognize smiles by measuring the geometry of the face, skin patterns, patterns of wrinkles, or changes in temperature associated with opening the mouth. Smiles are spontaneous, so the software must incorporate a vector field capable of following geometric distortions in real-time. The same programs can be applied to blink recognition, which prevents you from taking a picture of a subject with his or her eyes closed.

Computers can recognize smiles and other emotions, a feature that is thought to be key in the future acceptance of artificial intelligence by humans (we want machines to be sympathetic and kind to us). Because less than 10% of human communication is based on actual words, recognizing changes of facial expression is critical. Computers are able to recognize varying degrees of smiling on the basis of assigning them specific percentages. These sophisticated programs accomplish this by measuring not only the mouth but also pupillary reactions and wrinkling of the eyes. 3D systems allow recognition of smiles even when faces are turned or out of focus. A smile decrypting program was tried on the most baffling of smiles: the Mona Lisa. Researchers using sophisticated computer analysis say Mona is 83% happy, 9% disgusted, 6% fearful, and 2% angry.[3]

Emotion recognition programs work by tracking the place and movement of about 12 points in the human face. Face tracking algorithms match movements to 6 basic expression patterns: anger, sadness, fear, surprise, disgust, happiness, or a mixture of these. Companies that study marketing use these programs to measure customer likes and dislikes. The most honest reactions occur in privacy, so subjects are generally tested in empty rooms with hidden cameras. The Glad or Sad Web site analyzes pictures and rates them according to the 6 previously mentioned facial expression patterns.[4] Let's say that you took a picture of George Clooney but cannot read his expression. Simply upload it, and the program will analyze it (you are likely to find out that he was angry at having his portrait taken without permission).

Smiles can be improved as evidenced by many Web sites containing pictures of smiling celebrities before and after cosmetic dental treatments. When I go to the dentist, I am surprised (and somewhat intimidated) by a plethora of posters asking me if I am happy or ashamed of my smile (I am neither).

Some 20 years ago, orthodontists treated only children, but today 25%–50% of their work is done on adults. These are folks whose parents did not have the means to get their teeth straightened or individuals who stopped using their retainers before reaching 25 years of age (when the face finally stops growing). Straight teeth make cleaning easier and wear more evenly. Adult treatments take longer and are more painful, but overall they involve the same hardware used in children. Uncomplicated treatments start at about US $3000, and dentists are willing to treat even those individuals older than 90 years of age who desire to improve self-confidence by way of a beautiful smile. Some Northern European countries (such as the Scandinavian ones) offer orthodontics free of charge to their citizens younger than 18 years of age.

Dental braces have been around forever. Mummies dating back to 500 BCE harboring metal dental hardware in combination with natural animal fibers have been found in the Middle East and in the Mediterranean countries.[5] A French dentist, Pierre Fauchard, is credited as the father of modern orthodontics.

Later another French dentist established the practice of extracting the premolars to alleviate crowding and improve the growth of the mandible. For his work, he was named dentist to the King of France.

At the start of the 20th century, straightening crooked teeth took second place to correcting mandibular and malocclusion defects—that is, the alignment of teeth ceased to be the main goal, and correcting malocclusions became paramount to a successful treatment. In reality, one cannot be accomplished without the other. Malocclusions can be class I (neutrocclusion) when the relationship of the occlusion is normal for the maxillary first molars but not for the others. Class II (distocclusion) refers to anterior displacement of the upper molars and can be further divided if the anterior teeth protrude or are retroclined.[6] Crowding of teeth is generally due to a small maxilla or mandible, which, for unknown reasons, seems to occur predominantly in Western individuals. Of course, one of the complications of dental braces is that teeth may become discolored, a minor problem easily solved with bleaching.

Dental bleaching or whitening is perhaps the most common and benign procedure in the armamentarium of cosmetic dentistry. Our teeth become dark and yellow with age as the enamel changes and becomes (normally) stained. The most popular bleaching methods are strips or trays that contain oxidizing agents such as hydrogen peroxide or carbamide peroxide. If done at the dentist's office, a light energy source (generally halogen) may be used to accelerate the process (the same is found at many shopping malls, spas, and even in some of the more expensive home kits). Lightless treatments are called low-concentration and take longer and are less effective than high-concentration ones (which use light enhancement and higher concentrations of bleaching agent). Teeth stained by tetracycline take longer to whiten regardless of treatment type. Side effects are minimal and generally include only a temporary increase in tooth and gum sensitivity. High-concentration treatments may cause dehydration of the enamel, and the consequences of this are not known. Some have concerns regarding the potential carcinogenic effects of peroxides, but this has not been proved.[7] If bleaching fails, veneers can be applied for those who still desire a dazzling white smile.

Smiles generally mean happiness, and because we expect ourselves and those around us to be happy, we tend to smile a lot. Never has so much smiling been seen since the introduction of Happy Face (or Smiley) in 1963. Smiley used to be everywhere in the form of stickers, and now it has come back as an emoticon available on all e-mail systems and mobile telephones. Something that always baffles me is that we are expected to smile when asked how we are doing even if we are feeling terrible. A few days ago, I got into an elevator with a colleague of mine who asked me how I was feeling that morning.

Being the typical insensitive person that I am, I decided to respond with the truth, "Awful, did not sleep well last night, children problems at home, and in desperate need of a cup of coffee." Well, she certainly acted offended at my answer, so the next day when we took the elevator together again, with trepidation, she once more asked how I was doing. I decided to lie, "Wonderful, today life is just peachy," and flashed her a smile with my perfectly aligned ghost-white teeth. This time she responded with a beautiful smile equally full of white perfect teeth.

References

1. Strathearn L, Li J, Fonagy P, et al. **What's in a smile? Maternal brain responses to infant facial cues.** *Pediatrics* 2008;122:40–51
2. Kitts CD, Egan G, Gideon DA, eta al. **Dissociable neural pathways are involved in the recognition of emotion in static and dynamic facial expressions.** *Neuroimage* 2003;18:156–68
3. Bigger S. **Behind that smile.** *New Scientist* 2005;2530:25
4. Glad or Sad. http://www.gladorsad.com/en. Accessed September 28, 2011
5. Dental Braces. http://en.wikipedia.org/wiki/Dental_braces. Accessed September 28, 2011
6. Orthodontics. Angle's classification of malocclusion. http://web.archive.org/ web/20080213164657/http:/www.unc.edu/depts/appl_sci/ortho/introduction/ angles.html. Accessed September 28, 2011
7. Charlie Brooker. Thinking of getting you teeth whitened? Well don't. Keep them brown. *The Guardian*. November 12, 2006. http://www.guardian.co.uk/commentisfree/

HEART OF DARKNESS

For me, the word "evil" is somewhat of an abstraction. Having led a protected life, one of only a few occasions on which I have seen something resembling evil was when the Twin Towers fell on September 11th. Evil is the intent of harm and deliberate violation of basic moral and ethical codes. Evil is now woven into our culture, something that, in recent times, probably became more apparent (and popular) when President Reagan called the Soviets "evil" and, later, with Mr. Bush extending this concept to the "axis of evil." Invasion and domination by conquering forces are explicitly evil, as we have learned by how the Spanish (and we Americans) treated our native populations in the Americas. *Heart of Darkness* by Joseph Conrad recounts the evils that Belgians brought upon the Congo (the same story, transposed to our involvement in Vietnam, was used by Francis Ford Coppola in *Apocalypse Now*).

Conrad implied that darkness lies in all of our hearts and that we are all capable of evil.

Evil loomed over all of us during the cold war (but without overtly manifesting itself) and is now omnipresent in popular culture. Actor Mike Myers ridiculed it when he portrayed Dr.

Evil in the silly Austin Powers movies. Evil is a mostly male behavior and a rite of passage in many groups (i.e., gangs) and societies. The concept of evil is inseparable from religion; the forces of good and evil are always at battle with each other, particularly in our Western beliefs. Evil is human, and when we perceive some animal acts as such, it is just a projection of our own morals. In his book, *Evil: An Investigation,*[1] Lance Morrow speaks of micro and macro evil using as examples rape and genocide, respectively. Although evil resides in all of us, most well-balanced individuals control their urges and nothing comes of it. Evil can creep inadvertently into someone, as William Peter Blatty showed us in his book (and, later, in the movie directed by William Friedkin), *The Exorcist*. Evil can be highly efficient and bureaucratic, as with the Nazis or the Soviet Gulags or the killing of the Congolese by King Leopold II of Belgium (again portrayed in *Heart of Darkness*).

The word "evil" was probably coined during the Middle Ages from the Old English "yfel" and/or the German "ubil."

The common idea behind these 2 words is that of transgression.

[2] Many think that the idea of evil—as we now interpret it—was first mentioned in the Bible (*Genesis* 2:18), in which the Lord says, "It is not good for the man to be alone. I will make a helper suitable for him."[3] It is the idea of being alone that is evil, perhaps because if there is no other person who can check our evil thoughts, then nothing may prevent us from doing harm. In classic Greek philosophy, Plato had many opinions regarding evil and its relationship to ignorance, false words, and indifference to the public good.[4] Moral absolutists tell us that good and evil are fixed concepts. Amoralists or moral nihilists, such as Machiavelli and Stalin, claim that evil does not exist. Between these extremes, one finds the moral relativists who claim that evil is directly linked to local culture.

Others consider that people are not evil and that the term applies only to their acts. Islam does not recognize evil in the same way traditional Judeo-Christian theology does. All religions that began in Mesopotamia and adjacent territories, however, recognize evil in the form of Satan (Judaism: ha- Satan, Christianity: devil, and Islam: Shaitan). Satan was the chief of the fallen angels, that is, angels who were banished from Heaven and sent to earth to tempt man into committing evil. Lucifer (meaning "the shining one") is the most distinguished one and appeared after the fifth century of the Common Era. In the Quran, Iblis plays the role of Lucifer. But it does not matter who or what tempts us into evil—its results can be spun off in different ways according to circumstances and, nowadays, according to the media. But, as Mr. Morrow says in his book, you can dress evil as ethnic, political, psychiatric, and other excuses but it is still…just evil in costume.

Evil can be reduced to a person or group of persons. When studied with brain imaging, evil individuals have smaller amygdala volumes than controls.[5] Evil individuals clearly display antisocial behavior. Antisocial individuals have a significantly increased incidence of cavum septum pellucidum.[6] A study suggests that the presence of this cavum may reflect maldevelopment of the limbic system. To us in neuroradiology, it may be difficult to relate such a common anatomic variation with such extreme behavioral aberrations, but recent genetic studies start to shed some light on what is going on in the evil brain. Absence of a gene that encodes for monoamine oxidase A has been reported in antisocial human behavior.[7] Animal models that allow for temporary manipulation of this gene show that after turning it back on, the aggressive animals revert to normal behavior.[8] Studies have shown that up to 7 genes may be responsible for aggressive and antisocial behavior in humans. These observations are supported by studies of twins who, independent from their rearing, show genetic influences related to antisocial and aggressive behaviors. Raine et al[9] showed that antisocial individuals have an 11% reduction in prefrontal gray matter volume compared with controls.

These same authors and other investigators have shown that violent offenders have reduced glucose metabolism on PET in the same brain areas.[10] Prefrontal damage leads to developmental sociopathy rather than acquired sociopathy. Developmental sociopathy is closely associated with impaired moral reasoning and judgment.[10] Patients with lesions in the orbitofrontal cortex display a behavior characterized by lack of concern with the consequences of their acts, and aggressive individuals show less uptake of glucose on PET in those same regions. fMRI demonstrates abnormal activation of the same areas in impulsive individuals. The temporal lobes are equally affected, showing abnormal volumes, abnormal cerebral perfusion on SPECT, and reduced glucose metabolism on PET.[10]

Reduced activation on fMRI studies also has been demonstrated, though less consistently, in the parietal lobes and cingulate gyri of antisocial individuals.[10]

If evil is related to morality, where is the brain's moral center? Multiple studies show activation of the prefrontal cortex when making moral decisions. Using nuclear imaging and fMRI, individuals have been studied while viewing pictures depicting moral and immoral violations.[10] All showed activity changes in the prefrontal cortex and, less consistently, in the posterior cingulate, amygdala, and anterior temporal regions.

Moral judgments are directly related to emotion rather than reasoning and cognition. Antisocial individuals have the ability to reason between what is wrong or right but cannot apply it to their own behavior. Thus, they know, but lack a feeling for, what is moral. In them, the prefrontal cortex and amygdala do not work well. These areas are implicated in self-appraisal, self-reflection, self-perception, and insight. We know that in illegal drug use, child abuse, and child neglect, the prefrontal cortex and the limbic system suffer irreversible changes. If criminals are intellectually capable of distinguishing right from wrong, but are unable to emotionally control it, the argument of "not guilty by reason of insanity" seems to fall flat. Here is where medicine intersects law and the discipline of "neuroethics" was born.

Martha J. Farah, from the University of Pennsylvania, is considered, by many, as the "mother" of neuroethics.[11] This new discipline addresses ethical issues within the neurosciences.

What should be interesting to us is that neuroethics concerns fMRI and other methods of brain imaging. "Brain privacy" may be easily violated by imaging, as information obtained may be the result of a study designed for other purposes.

Lie detection has been used in court, and something called "brain fingerprinting" is now considered an adequate method of screening for terrorists* and is accepted in court.[12]

"Brainotyping" reveals that certain individuals are predisposed to violent crimes, pessimism, risk aversion, unconscious racial attitudes, and even differences in sexual attraction! Of course, the media would have you believe that a scan can be done and, presto, you know an individual's personality. This is infinitely more complex and requires intense psychiatric evaluations.

Because individuals with aggressive and antisocial behaviors tend to demonstrate relatively specific localizations of abnormal brain morphology and function, it seems that disconnecting these areas or modulating their functions may result in behavioral changes and, hopefully, improvement. Contemporary psychosurgery began in the mid-1930s by disconnecting the prefrontal cortex by leucotomy. For neuroradiologists, it is important to remember that the "inventor" of cerebral angiography, Edgaz Moniz, received the Nobel Prize not for angiography but for his work on frontal lobotomies, either mechanical or by alcohol ablation. In the 1950s, the use of lobotomy was replaced by chlorpromazine.[13] Limbic leucotomy seems to have beneficial effects in cases of self-mutilation and repeated assaultive behavior. Today, psychosurgery is a well-recognized field and procedures commonly involve ablations, which are permanent compared with the more flexible neuromodulation via stimulators. The benefit of stimulation is that it can be turned off, if needed. How neuromodulation can be used in criminals is a matter of speculation.

Are we, as a society, more evil than ever before? If hate is a measure of evil, we are certainly heading that way. The definition of the word hate includes extreme hostility toward something or someone we dislike. Hate can be verbal, as in hate speech, or physical, as in hate crime. In February 2011, the Southern Poverty Law Center issued a report stating that there are now over 1000 hate groups in the United States (more than ever before).[14] Most of these individuals hate their government, as they feel that it is depriving them of freedom. Hate groups are generally involved in violent criminal acts. Because we need to defend ourselves from evil, what could be more logical than buying a gun? On the same week as the previous report was made public, the state of Texas seemed poised to pass a law permitting concealed weapons on university campuses.[15]

The evidence supports the notion that evil is increasing and, as neuroradiologists, we may be called upon to assess it with our sophisticated imaging techniques. After all, it seems that darkness does not reside in the heart but rather in specific locations in the brain.

*Factoid: After the French revolution, Robespierre was in charge of the "reign of terror." During this time, thousands were killed by his supporters, who were known as "terrorists." Thus, terrorists are persons who instill fear and kill indiscriminately (basically, evil individuals).

REFERENCES

1. Morrow L. *Evil: An Investigation.* New York: Basic Books,; 2003
2. http://www.merriam-webster.com/dictionary/evil. Accessed November 15, 2011
3. http://www.biblegateway.com/passage/?search_Genesis_2%3A18& version_NIV. Accessed November 15, 2011
4. http://www.quotationspage.com/quotes/Plato. Accessed November 15, 2011
5. http://bigthink.com/ideas/23882. Accessed November 15, 2011
6. Raine A, Lee L, Yang Y, et al. **Neurodevelopmental marker for limbic maldevelopment in antisocial personality disorder and psychopathy.** *Br J Psych* 2010;197:186–92
7. Raine A. **From genes to brain to antisocial behavior.** *Curr Dir Psychol Sci* 2008;17:323–28
8. Cases O, Seif I, Grismby J, et al. **Aggressive behavior and altered amounts of brain serotonin and norepinephrine in mice lacking MAOA**. *Science* 1995;268:1763–66
9. Raine A, Lencz T, Bihrle S, et al. **Reduced prefrontal gray matter volume and reduced autonomic activity in antisocial personality disorder.** *Arch General Psych* 2000;57:119–27
10. Raine A, Buchsbaum M, LaCasse L. **Brain abnormalities in murderers indicated by positron emission tomography.** *Biol Psych* 1997;42: 495–508
11. http://www.psych.upenn.edu/_mfarah/index.html. Accessed November 15, 2011
12. www.brainwavescience.com. Accessed November 15, 2011
13. Mashour GA, Walker EE, Martua RL. **Psychosurgery: past, present, and future.** *Brain Res Rev* 2005;48:409–19
14. http://www.npr.org/2011/02/23/133970226/new-report-higher-hate-groupcount- than-ever. Accessed November 15, 2011
15. http://www.dailytexanonline.com/news/2011/05. Accessed November 15, 2011

SOME THINGS ARE BETTER LEFT UNSAID

> *My ideas seem to frighten you . . . some things are better left unsaid.*
>
> *Daryl Hall and John Oates*

In law enforcement, as in radiology, it is customary to ask a witness (in our case our trainees) to repeatedly describe an observed event (in our situation the imaging findings). Sometimes we ask our trainees to describe the features of different lesions even when they are not looking at images. This is based on our belief that repeated verbal processing builds mental images that we can all later retrieve in an attempt to match them with what we are seeing on our studies. We all have been taught this way, and because of this, we think it is a good method and continue to use it, but, are we doing the right thing?

In 1990, two psychologists from the Universities of Pittsburgh and Washington noticed that previous verbal articulation distorted future visual recognition, a phenomenon they named "verbal overshadowing."

[1] For their dissertations, they tested several hypotheses, one being that verbalizing the appearance of something previously seen impaired its future recognition (this is exactly what I mistakenly ask my residents to do: "Describe for me the imaging findings of the glioblastoma we saw last week"). Schooler* and Engstler-Schooler did a series of simple tests, and one went like this: The participants were shown a face; some were asked to repeatedly describe that face and a control group was not asked anything. Later the same face was shown to all participants—guess who did better? The rehearsed group did much worse! It seems that the brain has specific systems that help it remember faces, a common task needed for socialization.

A well-known condition that impairs this task is prosopagnosia, in which subjects have varying inabilities to recognize faces. More interesting, our brain recognizes faces more easily when they are in the usual orientation. If faces are presented upside down, we have trouble recognizing them. Repeated verbalization also decreases the chances of correctly recognizing a face. Further experiments determined that when visual stimuli are difficult to verbalize, the memory of them is actually impaired instead of improved by repeated verbalization. Perception of color is also affected by verbal descriptions. When shown colors and then asked to describe them, subjects ended up choosing the one that best fit their description rather than the one they were initially shown.

This makes me wonder if persistently asking our trainees to describe the signal intensity of lesions may actually be detrimental to future information retrieval and interpretation of images.

What about our trainees reporting the same findings over and over again by dictating them? Does this improve their (our) memory?

What we radiologists do is to verbalize something that is nonverbal: our perception of images. It is known that verbalizing stimuli facilitates memory but only if the stimuli match our ability to use language to describe them; in this way, both effects are concordant and additive, and memory is facilitated. Thus, rehearsing something like a written history lesson may help us do better on a written examination because in this situation there are no images, only words. In a different experiment, subjects were given a simple verbal statement, and some were asked to repeat it out loud while others were not. In this situation, memory retrieval was improved in those who verbalized the statement, but curiously the improvement was only marginal. Could it be that dictating the same findings innumerable times during our lifetimes as radiologists only minimally helps us to be better professionals?

Not only does verbalizing interfere with immediate memory recall but it affects our long-term memory. There is, however, hope for radiologists: By limiting the time we need to recognize an image, the effects of verbal overshadowing decrease. Therefore, it may be better to interpret cases quickly, rather than taking too much time to think about them! As our schedules become more complicated and full, we spend less time interpreting images, and it is possible that this paradoxically increases our accuracy rather than decreasing it as commonly thought. It is also true that waiting longer between verbalization and re-evaluation of images reduces the effects of overshadowing.[2] Both Schooler and Engstler-Schooler went on to publish several more articles that confirmed their initial observations, which they named the "recoding interference hypothesis." This hypothesis basically states that verbalizing a visual memory produces a biased memory representation that interferes with the original visual memory and its future recognition. To me, the implications of these observations for radiologists are staggering and deserve a closer look.

In 2001, two different investigators published a meta-analysis of 29 verbal overshadowing experiments performed in 2000 participants.

[3] This analysis included only studies dealing with face recognition, and the authors concluded that the data clearly indicated that subjects who described a face were much more likely to misidentify it subsequently compared with those who did not generate a description before identification. It appears that we self-generate misinformation and manipulate its output, leading the subsequent recognition errors.[4] To complicate things even further, if participants in a study are given elaborate and detailed instructions as to what to recall, they show poorer results than those who are given the right to free recall. Again, as it pertains to radiology, are we harming ourselves by

demanding from our trainees highly structured and detailed reports for each study they interpret? Would a more accurate representation of imaging findings occur in free-form reports?

Evidence exists that verbal overshadowing extends beyond face and color recognition to also affect wine tasting, decision-making, voice recognition, and insight problem-solving.[4] An extension of the face recognition studies looked at the influence of verbal overshadowing with respect to attractiveness.[5] These investigators found that individual perception of attractiveness was highly influenced by the amount of verbal attention it received. This influence was most prominently found in females compared with males. It seems that women can change their perception of human face attractiveness by verbalizing. In my experience, this makes sense because we men are more rigid in our perception of beauty and rarely change our initial impressions of it. Verbal overshadowing extends to taste, and as such, food (and wine) generally does not taste exactly the same way it did the first time we tried it. I know that after tasting something I really liked and talking about it for weeks, the second taste is always surprisingly different and many times disappointing, though this could be due to differences in preparations.

Because visual memory representations incorporate visual, spatial, and temporal characteristics, studying a set of images (such as CT/MR images consecutively displayed in sets of 6, 9, or 12 images on each monitor) is very different from studying individual images presented in sequence (such as "stacking" CT/MR images in the monitor).

Learning sequences of images recruits the verbal system (and thus verbal overshadowing); something that does not occur with individual pictures. So, it is possible that looking at images the way they were traditionally displayed on film (sequentially) is counterproductive.

I consider myself a good clinical neuroradiologist but hate looking at images that are stacked. Regardless of that, my interpretations are generally correct, consistent, and fast. After thinking about it, I realize that I do not look at images sequentially but look at the 9 images per monitor at the same time as if they were a single picture.

This allows me to be very fast, and this speed in combination with the non-sequential evaluation may decrease my verbal overshadowing and permit me to recognize abnormalities faster and easier. Additionally, I never try to convince myself that I know what the studies show; and because trainees dictate all of my reports, the effects of verbal overshadowing are minimized. Investigations have shown that if one separates the visual from the spatial from the temporal characteristics in an image, the effects of verbal overshadowing are prevented or attenuated.[6] It seems that our brains retain images and verbal representations at different rates, and both are available to use differently also. Because most research has been done by using static images (pictures), some argue that these do not adequately replicate reality that is characterized by fluidity (thus using videos may be a better choice).[7]

Environmental stimuli are of 2 types: static and dynamic. I have already discussed the effect of verbal overshadowing on static images, so I will now concentrate on events that happen in a dynamic fashion. Events consist of objects in situations characterized by a constant change in spatial relations, mainly distances and orientations, with time.[8] While verbalization of static images may be considered as concrete, verbalization of events is abstract.

If we put an event into words, we end up with 2 competing models: a verbal description and an observation. These 2 models interfere with each other, and that is why events never occurred the way we describe them initially. Not only are our initial verbal descriptions different from the event itself, but subsequent verbal descriptions are different from the initial ones. People who were asked to write down what they saw on September 11, 2001, were asked again several years later to rewrite their impressions, and the 2 versions were so different that many complained that what was given to them as their initial version was false. The reverse is true:

You see what you want to see. Individuals asked to verbally describe an unseen event will shape their perception of it by recognizing mostly the parts previously described (these parts are said to become "prioritized"). This may not be all that bad. Verbal before- event descriptions help us to identify distractors, allowing us to concentrate on what is important. This observation applies only to dynamic events and does not improve recognition of static images. Thus, teaching our trainees what is important to look for in static images is probably not beneficial; however, teaching them what to look for in dynamic studies such as cerebral angiograms is very useful because it filters out the "noise."

How can we attenuate the effects of verbal overshadowing? All of us have seen (on television, at the movies) hypnosis used by law enforcement agencies (and spies) to increase recall. If one can eliminate the tainted consciousness, perhaps tapping the unconscious will result in better recollections. Studies have looked at this issue and show that the opposite is true; that is, hypnosis actually decreases accuracy, leads to false confidence, and increases suggestibility leading to the procurement of misleading information.

[9] Solutions to verbal overshadowing may be simpler: Avoid too much verbalization, after verbalization engage in a nonverbal task, increase the length of time between verbalization and image recognition, try to verbalize your impression of dynamic events, use free-form verbal expressions rather than highly structured ones, and avoid describing colors.[10] Another observation that I find of

tremendous importance to us is the effect of overshadowing on the type of recall required. It seems that if you ask individuals for "piecemeal" descriptions and recollections, overshadowing of subsequent recognition is much greater than if you ask them for "elaborative" descriptions. This brings me to the last part of this essay: Most books from which we study nowadays present information in a bulleted or piecemeal fashion rather than a long prose form. Because subsequent image discrimination is more affected by previous elaboration of piecemeal descriptions, are we still learning the right way? Would it be better to go back and read books that present information in an elaborative fashion?

Okay, maybe I am exaggerating, but I am trying to play the devil's advocate by bringing up these issues regarding verbal overshadowing.

It just seems that they are important enough that perhaps further investigation into how they apply to radiology may be needed. Next month's *Perspectives* will deal with thoughts about the use and misuse of the scientific process.

* Dr. Jonathan Schooler is now Professor of Psychology at the University of California in Santa Barbara. His original article (1990) on verbal overshadowing has been quoted more than 400 times in the scientific literature.

REFERENCES

1. Schooler JW, Engstler-Schooler TY. **Verbal overshadowing of visual memories: some things are better left unsaid.** *Cogn Psychol* 1990;22:36–71
2. Finger K, Pezdek K. **The effect of the cognitive interview on face identification accuracy: release from verbal overshadowing.** *J Appl Psych* 1999;84:340–48

3. Meissner CA, Brigham JC. **A meta-analysis of the verbal overshadowing effect in face identification.** *Applied Cog Psych* 2001;15:603–16
4. Meissner CA, Memon A. **Verbal overshadowing: a special issue exploring theoretical and applied issues.** *Applied Cog Psych* 2002;16:869–72
5. Talbot BH, Gifford JL, Peterson E, et al. **The verbal overshadowing effect: influence on perception.** *Intuition* 2008;4:12–18
6. Pellizon L, Brandimonte MA, Luccio R. **The role of visual, spatial, and temporal cues in attenuating verbal overshadowing.** *Applied Cog Psych* 2002;16:947–61
7. Meissner CA, Brigham JC, Kelley CM. **The influence of retrieval processes in verbal overshadowing.** *Mem Cognit* 2001;29:176–86
8. Huff M, Schwan S. **Verbalizing events: overshadowing or facilitation?** *Mem Cognit* 2008;36:392–402
9. Kebbell MR, Wagstaff GF. **Hypnotic interviewing: the best way to interview eyewitnesses?** *Behavio Behav Sci Law* 1998;16:115–29
10. Brown C, Lloyd-Jones TJ. **Verbal overshadowing in a multiple face presentation paradigm: effects of description instruction.** *Applied Cog Psych* 2002;16:873–85

THE SIXTH DIMENSION AND GOD'S HELMET

I have to be honest: my feelings about religion are ambivalent. The famous author Julian Barnes expresses similar feelings at the start of his book *Nothing to be Frightened Of*: "I don't believe in God, but I miss him."[1] I guess that calling oneself an agnostic may be considered a wishy-washy position by many but that seems to fit his position (and mine).

Agnosticism: the view that the truth value of certain claims—especially claims about the existence or nonexistence of any deity, but also other religious and metaphysical claims —is unknown or unknowable.[2] A more decisive individual may call him or herself an atheist. Mr. Barnes tells us that category 1 atheists are those who have no God and no fear of death. Despite this, most still enjoy life and show wonder at our lives and world.

Atheism: (opposite of theism) in a broad sense, is the rejection of belief in the existence of deities. Atheism is simply the absence of belief that any deities exist.[3]

Is the human predisposition to religion cultural or biologic?

Studies suggest that activation of certain cerebral networks spanning the frontal, parietal, and temporal lobes are associated with spiritual states. Thus, damage to these areas should alter some of these feelings, including self-transcendence.

Transcendence refers to the ability to detach one's consciousness from the physical body, a type of spiritual experience. Transcendence is common to all faiths and is part of the so-called "religious state." Spiritual experiences are as common today as in the past. Catholic nuns and Buddhist monks undergoing functional MR imaging (fMRI) during introspection showed changes in the prefrontal, cingulated cortex, temporal and parietal lobes, and in some subcortical areas.[4-8] These brain changes varied with the individual's ability to meditate. The activity of the serotonin brain system is probably genetically determined and is linked to varying degrees of self-transcendence and religious or spiritual experiences.[9]

In an interesting study, Urgesi et al[10] obtained self-transcendence scores before and after brain tumor resections.

They postulated that selective surgical damage to the frontal lobes would decrease self-transcendence, whereas temporoparietal damage would increase it. They tested nearly 90 patients with different tumors (high- and low-grade gliomas, meningiomas) and found that when posterior areas of the brain were removed, a significant and reliable increase in self-transcendence occurred. This change happened very soon after surgery and therefore was not considered an adaptive process. Furthermore, patients in whom meningiomas were removed from similar locations did not experience these changes (the underlying brain was presumably not damaged as consequence of an extra-axial mass resection). Patients who were already highly religious before surgery reported increased self-transcendence and mystic experiences postoperatively.

It is not clear from this study whether damage to posterior brain tissues lead to higher recruitment of activity in other regions of the brain that contribute to the feeling of transcendence. In a different study, 12 patients (6 religious) were evaluated with positron-emission tomography imaging while praying.[10] From memory, they recited Psalm 23 while undergoing cerebral blood flow studies. For reasons that escape me, Psalm 23 is usually a favorite of religious converts. In this study, all religious subjects reported having attained a "religious state" during recitation and showed significantly increased blood flow to the right dorsolateral prefrontal cortex.

This contradicts the common belief that it is mainly the limbic system that is associated with faith beliefs. Religious beliefs may be a cognitive process mediated by pre-established neural circuits.

Could it be that God is located in one part of the brain?

Beauregard and Paquette[5] performed fMRI in a series of Carmelite nuns while they reported being in a state of union with God. The Carmelite Sisters are the counterpart of the Carmelite Brothers, and both groups follow very strict dietary (could this influence brain function?) and religious beliefs and activities. The 15 Carmelites evaluated showed positive fMRI results but instead of activating only one region, 12 different regions were activated. Franciscan nuns seem to activate slightly different regions of their brains when praying.[11]

In both groups, the pre- and inferior-frontal regions were active during the tasks performed. Because spiritual and mystical experiences are relatively common in patients with temporal lobe seizures, microseizures are long thought to be responsible for some of these experiences.

The fact that epileptics have numinous experiences more often than non-epileptics may explain why they have been revered in some cultures and persecuted in others. Sigmund Freud dismissed spiritual experiences as pathology, but not surprisingly Carl Jung did not. The "Sacred Disease" of antiquity was refuted by Hippocrates who argued against the association of seizures and prophetic and mystical powers (for him seizures were just purely a brain dysfunction). In the New Testament, Matthew (17:14 –20) witnessed Jesus curing a boy who presumably had epilepsy and this may be one source from which the association between epilepsy and religion comes. In an article reviewed for this essay, I found a fascinating table listing religious figures who allegedly had epilepsy, including the Buddha, Mohammed, Ezekiel, St. Paul, and Joseph Smith among others.[12] Before and after seizures, up to 4% of patients report having religious experiences.[12] Patients with postictal psychosis may experience feelings of hyper religiosity (Joan of Arc may have been one such patient).

Religion has been present throughout the history of humankind and some argue that humans should be called *Homo religious* rather than *Homo sapiens*.[13] More than one-half of American adults report a "spiritual" and life-changing experience.

That includes people with many different personalities and backgrounds. Personality researchers propose 5 main dimensions of personality: extraversion, agreeableness, conscientiousness, neuroticism, and openness. These 5 traits probably define personality but need not all be present in one person. Religiousness is probably the sixth major dimension of personality.[13]

If religious experiences are based on neurophysiologic events, can we provoke them? In 2001, Michael Persinger reported using an apparatus to stimulate spiritual experiences.[14]

Dr. Persinger is a well-recognized personality in the world of paranormal studies. His research extends to the perception of unidentified flying objects (UFOs). He states that UFOs and other paranormal experiences can be caused by changes in the magnetic environments in which we live (though I have never heard a patient say anything about paranormal experiences during or after an MR imaging at 1.5T or 3T). Based on his observations, he outfitted a snowmobile helmet (and if you Google it, that is exactly what it is) with solenoids that presumably altered the local magnetic fields (as in transcranial magnetic stimulation). Compared with those we use in neuroradiology, his magnetic fields were miniscule and on the order of 3–7 microtesla. He reported that at least 80% of participants felt a presence in the room when wearing the activated apparatus.*

Some even categorized this presence as being God. Dr.

Persinger also seems to think that spiritual activity is located in the right temporal lobe, contradicting the evidence presented previously in this *Perspectives*. In 2004, a group of Swedish researchers tried to reproduce his results without any luck.[15] A significant number of subjects in the control group also experienced a presence, sometimes a significant one. As expected, prolonged and not so amicable arguments erupted between them and Dr. Persinger. Back-and-forth arguments included inadequate exposure to magnetic waves due to a short time wearing the helmet and the different and varying degrees of suggestibility in the subjects.

To prove his point, he chose to try this thing on Dr. Richard Dawkins. Because Dr. Dawkins—Professor of Public Understanding of Science at Oxford—is a world-class category 1 atheist, this experiment can be interpreted as very brave or very foolish on Dr. Persinger's part. Not unexpectedly, Dr.

Dawkins did not feel a presence or any effects while wearing the contraption. Just for the record, popular publications by Dr. Dawkins include *The Blind Watchmaker* (arguing against creationism and intelligent design), *The God Delusion* (faith is a delusion), and *The Selfish Gene* (gene-centered view of evolution), among others; thus, clearly, he is not a suggestible person. This experiment has not dissuaded others who continue to claim validity to the God helmet experiment. There are controversial data suggesting that alterations in the *VMAT*[2] gene (also involved in serotonin regulation) may affect an individual's degree of spirituality.[16] Will the helmet work better on these individuals? Neurotheology is a new discipline that attempts to answer many of these questions with the use of functional neuroimaging.

If the brain's serotonin system is implicated in religious experiences, can certain chemicals be used to duplicate these experiences? Some drugs, mainly LSD, mescaline, ayahuasca, and peyote, are known to have chemical features similar to serotonin and share its receptors. In the 1950s and 1960s, LSD was given to terminal lung cancer patients, and most reported having an increased acceptance of death, because while on their drug "trip" they had mystical experiences that lead to the firm belief that life continues after death. There are several fairly credible studies, some performed at Harvard on Harvard students, using serotonin-like drugs to incite mystical experiences, all with positive results.[17] Certain mushrooms and cacti (such as peyote) also contain psychogenic compounds that induce joyful, mystical, and religious experiences (mushrooms and cacti have been used for this purpose since ancient times in the Americas). Psilocybin is one such drug, and Johns Hopkins recruits cancer patients to take this drug and evaluate the spiritual changes it may bring.[18] These "spirit-facilitating" drugs may offer some solace to terminal cancer patients.

Ingesting psilocybin-containing mushrooms was the topic of a series of books in the 1970s by Carlos Castañeda. Many of us who were teenagers during the 1970s avidly read this anthropologist's series of books dealing with shamanism, attempting to gain some knowledge of spirituality. A wonderful and highly recommended book on the LSD experience is *The Electric Kool-Aid Acid Test* by Tom Wolfe.[19]

Writing about religion-related issues is always treacherous and bound to upset some. Rest assured that I am respectful of all faiths and beliefs. As we get older, our thoughts relating to death and some faith implications arise often, and so I thought to give a short overview about how these relate to neuroimaging.

It is important to keep in mind that many American academics have strong spiritual beliefs. Approximately 40% of scientists and 7% of members of the National Academy of Sciences believe in a God and nearly 40% believe in human immortality.[20]

* Maybe just covering one's head is enough to experience some type of spiritual experience. Is that why most religious persons, regardless of faith, wear headgear?

REFERENCES

1. Barnes J. **Nothing to be Frightened Of.** New York: Vintage International; 2008
2. http://en.wikipedia.org/wiki/Agnosticism. Accessed October 22, 2010

3. http://en.wikipedia.org/wiki/Atheism. Accessed October 22, 2010
4. Azari NP, Nickel J, Wunderlich G, et al. **Neural correlates of religious experience.** *Eur J Neurosci* 2001;13:1649–52
5. Beauregard M and Paquette V. **Neural correlates of a mystical experience in Carmelite nuns.** *Neurosci Lett* 2006;405:186–90
6. Brefczynski-Lewis JA, Lutz A, Schaefer HS, et al. **Neural correlates of attentional expertise in long-term meditation practitioners.** *Proc Natl Acad Sci U S A* 2007;104:11483–88
7. Cahn BR and Polich J. **Meditation states and traits: EEG, ERP, and neuroimaging studies.** *Psychol Bull* 2006;132:180–211
8. Newberg A, Alavi A, Baime M, et al. **The measurement of regional cerebral blood flow during the complex cognitive task of meditation: a preliminary SPECT study.** *Psychiatry Res Neuroimaging* 2001;106:113–22
9. Borg J, Andre´e B, Soderstrom H, et al. **The serotonin system and spiritual experiences.** *Am J Psychiatry* 2003;160:1965–69
10. Urgesi C, Aglioti SM, Skrap M, et al. **The spiritual brain: selective cortical lesions modulate human self-transcendence.** *Neuron* 2010;65: 309–19
11. Newberg A, Pourdehnad M, Alavi A, et al. **Cerebral blood flow during meditative prayer: preliminary findings and methodological issues.** *Percept Mot Skills* 2003;97:625–30
12. Devinsky O and Lai G. **Spirituality and religion in epilepsy.** *Epilepsy Behav* 2008;12:636–43
13. Emmons RA and Paloutzian RF. **The psychology of religion.** *Annu Rev Psychol* 2003;54:377–402
14. PersingerMA.**The neuropsychiatry of paranormal experiences.** *J Neuropsychiatry Clin Neurosci* 2001;13:515–24
15. http://www.nature.com/news/2004/041206/full/news041206 –10.html. Accessed October 22, 2010
16. Hammer DH. **The God Gene.** New York: Doubleday; 2004

17. http://www.luminist.org/archives/7tongues.htm. Accessed October 22, 2010
18. http://www.bpru.org/cancer-studies/study-info.html. Accessed October 22, 2010
19. Wolfe T. **The Electric Kool-Aid Acid Test.** NewYork: Farrar, Straus and Giroux; 1969
20. Larson EL, Witham L. **Scientists are still keeping the faith.** *Nature* 1997; 386:435–36

THE INS AND OUTS OF SEXUAL IMAGING

Golf and sex are the only two things you don't have to be good at to enjoy.

Kevin Costner in *Tin Cup*

What happens inside the human body during sexual intercourse has fascinated artists, scientists, and the public since the start of humanity. Leonardo da Vinci (1452–1519) drew the internal anatomy of a couple engaged in intercourse in "The Copulation" and suggested that connections between the penis, the distal spinal cord, and the brain existed. In the Middle Ages and the Renaissance, the act of sex was not scientifically studied. The term "sexology" was first used in Victorian times to describe the relationship between men and women.[1] The idea of "sexual science" originated in Germany and Italy, and the term "sexual medicine" was not established until the 1970s. In the United States, the first physician who worked full-time in "diseases" of a sexual nature was Harry Benjamin, who studied mostly trans-sexualism.

Perhaps one of the most common names associated with sexual medicine is Alfred Kinsey, who collected the sexual clinical histories of more than 18,000 patients (a story entertainingly told by T.C. Boyle in his novel *The Inner Circle,* Viking Adult, 2003).

Kinsey was the first to quantitate a few aspects of sexual behavior.

His legacy lives on at the Kinsey Institute (http://www. kinsey-institute.org) at Indiana University in Bloomington. A few years after the apparition of the "Kinsey Report,"[2] Masters and Johnson began their studies of sexuality, all of which culminated with the creation of the International Society for Sexual Medicine, which now publishes the *Journal of Sexual Medicine* (Impact Factor: 3.55). Today, sexual medicine is a serious and mature specialty that incorporates physicians, psychiatrists, psychologists, and even neuroradiologists.

Sex has 2 goals: foremost reproduction and then pleasure. The main brain region controlling pleasure and thus orgasm is the medial preoptic area of the hypothalamus. In animals, stimulation of this region elicits pleasurable sexual responses via sympathetic (by way of the paraventricular hypothalamic nucleus) and parasympathetic mechanisms.[1] Axons from this region connect with the nucleus paragingantocellularis in the ventrolateral medulla, and axons from the latter extend down to the tip of the spinal cord (here it is curious to note that during the Renaissance, semen was thought to come down from the brain traveling by way of a canal in the spine and following a path that somewhat matches that of these axons). By way of the periaqueductal gray matter, the medial preoptic area may inhibit the hypothalamic nucleus, terminating the sensations of pleasure.

Direct observations of what occurs at the penis-vagina interface during sex were initially made by Masters and Johnson, who studied the mechanics of this activity by using an artificial penis (a funny episode in the popular TV series "Masters of Sex" recounts the experiments done with this instrument). Sonography was first used to study copulation in 1999, but the images obtained were of poor quality by today's standards.[3] Later in 1999, MRimaging was used to study the state of the female genital and pelvic organs during sexual arousal in 8 couples during 13 "encounters."[4] In this study, in which one author was a radiologist, the main conclusion was that local responses were similar for pre- and postmenopausal women. The article contains no MR images of the act of coitus. In 2001, Faix et al[5] asked a couple to have sex in the MR imaging unit (in face-to-face or "missionary" position). The main observation, among many uninteresting measurements that were perhaps used to justify publishing such an article, was that in this position the contact between the penis and vagina occurs mostly at the anterior cul-de-sac and anterior vaginal wall. One year later the same authors asked a couple (the same one?) to have intercourse in the MR imaging unit again, also in missionary position, and images showed that the internal contact between the organs was different and occurred along the posterior cul-de-sac and posterior vaginal wall.[6] The images illustrating those articles are not too different from the original da Vinci drawing previously mentioned. The conclusions of this latter study were simply silly:

"Initially the aim of the study was to 'copy' the genius of Leonardo da Vinci. We showed that an MR imaging scan of sexual intercourse in two positions is feasible and artistic but not as artistic as the images drawn by da Vinci."[6] I wonder whether today any reasonable scientific journal would publish a study with these conclusions!

MR imaging has also been used to map the spread of contraceptive gel during both simulated and real intercourse.[7] These experiments offer important information regarding the spread of intravaginal medications and how it relates to dosage (small or large doses do not appear to make a difference). Studies have found that gels spread evenly, covering the cervix and proving a good contraceptive barrier.

What happens during sex at a local level is, from a neuroradiologist's point of view, not too interesting and for purposes of this *Perspectives*, I will now attempt to summarize some of what presumably happens in our brains. Using fMRI, Meston et al[8] studied 6 women while being shown neutral and erotic films, and all reported moderate arousal to the latter. Areas that showed activation during sexual arousal included the inferior temporal lobes, anterior cingulate gyri, insular cortex, corpus callosum, thalami, caudate nuclei, globus pallidi, and inferior frontal lobes. In women, the activation was bilateral, while in men it was unilateral (mostly right-sided).[9] A different study in a larger number of men and women showed activations in the same regions and, in addition, in the medial prefrontal cortex, occipitotemporal cortex, and amygdalae, but only men showed significant activation of the hypothalamus.[10] Some of the cortical areas were also activated with other emotional stimuli and presentation of rewards. Thus, it seems that female and male brains respond to sex differently (to me, not surprising).

In males, stimuli leading to initial genital responses activate the left frontal operculum, probably related to imagining the forthcoming sexual act.[10] Then, anticipatory motor and somato- sensory imagery activates the left supramarginal gyrus and Brodmann area 2, respectively. A positive feedback mechanism consequently increases the response of these brain areas, but the frontal lobes also have the ability to disrupt these pathways and stop related brain activations. During orgasm, prominent activation has been shown in the paraventricular hypothalamic nuclei, periaqueductal gray matter, hippocampus, and cerebellar cortex, confirming observations made previously in animal models. Stimulation of the posterior pituitary lobe with release of hormones such as oxytocin also occurs during orgasm in both animals and humans.

Because fMRI can shed light on what happens during normal sex, it has also been used to evaluate deviant sexual behaviors. It is curious (and somewhat scary) that pedophiles activate the same brain regions, albeit more intensively, than non-pedophile controls. [11],[12] It seems that in sexually addicted individuals, the frontal brain loses its ability to stop the positive feedback mechanisms that recruit other brain regions, leading to widespread activations. [13] DTI has also shown low diffusivity in the superior frontal cortex in sexually addicted patients (similar results have been found in other behaviors with abnormal impulsivity such as kleptomania, gambling, and eating).[13] These observations suggest that neurons and axons are abnormal or abnormally organized in these regions in addicted individuals.

Because sex and love are related (not always, but fairly commonly), it is useful to explore what happens to the brain during love. There are 2 types of love: passionate (being in love with someone) and companionate (loving someone). The former is more intense and though both types occur simultaneously early in a relationship, they are different feelings that may not survive together in long-term unions (with time, passionate love becomes companionate). Passionate love is characterized by activations in all dopaminergic brain regions and decreased activity of brain areas associated with anxiety and fear.[14] Most studies have shown that the amygdalae tend to be deactivated during love, sex, and orgasm, decreasing fear and anxiety and contributing to the feeling of well-being experienced with love, sex, and orgasm.[15] Conversely, in deviant sexual behaviors such as pedophilia, the amygdalae activate abnormally.[12] The brains of pedophiles appear to be more severely impaired than those of other sexual offenders, including rapists, and some brain regions (such as the basal frontotemporal ones) are affected in these and other delinquents and criminals of a nonsexual nature.

Romantic love is also ruled by the dopaminergic systems of the brain. Love is primarily motivational and changes our habits to please the person who receives it. Central features that both love and sex share are intrusiveness and obsessive thinking, which are involuntary and difficult to control. In individuals sustaining long-term partners, the initial reward-value brain circuits remain activated similar to new love. Serotonin production is elevated with certain antidepressants, and this chemical must be present to feel love, so there is a chance that individuals taking these drugs may be unable to fall in love or have a tendency to fall out of it.

So, it seems that the mysteries of love and lust are starting to become clearer with fMRI. Although the study of sex has gone in and out of style many times, thanks to neuroimaging, it is definitively in now.

REFERENCES

1. Schultheiss D, Glina S. **Highlights from the history of sexual medicine.** *J Sex Med* 2010;7:2031–43
2. Kinsey AC. Pomeroy WB, Martin CE, et al. *Sexual Behavior in the Human Female.* 1st ed. Philadelphia: Saunders; 1953 (still available on Amazon.com)
3. Riley AJ, Less W, Riley EJ. **An ultrasound study of human coitus.** In: Bezemer W, Cohen-Kettenis P, Salob K, et al, eds. *Sex Matters.* Amsterdam: Elsevier; 1999:29–36
4. Schultz WW, van Andel P, Sabelis I, et al. **Magnetic resonance imaging of male and female genitals during coitus and female sexual arousal.** *BMJ* 1999;319:18–25

5. Faix A, Lapray JF, Courtieu C, et al. **Magnetic resonance imaging of sexual intercourse: initial experience.** *J Sex Marital Ther* 2001; 27:475–82
6. Faix A. Lapray JF, Callede O, et al. **Magnetic resonance imaging (MRI) of sexual intercourse: second experience in missionary position and initial experience in posterior position.** *J Sex Marital Ther* 2002;28:63–76
7. Pretorius ES, Timbers K, Malamud D, et al. **Magnetic resonance imaging to determine the distribution of a vaginal gel: before, during, and after both simulated and real intercourse.** *Contraception* 2002;66:443–51
8. Meston CM, Levin RJ, Sipski ML, et al. **Women's orgasm.** *Annu Rev Sex Res* 2004;15:796–97
9. Stoleru S, Gregoire MC, Gerard D, et al. **Neuroanatomical correlates of visually evoked sexual arousal in human males.** *Arch Sex Behav* 1999;28:1–12
10. Karama S, Lecours AR, Leroux JM, et al. **Areas of brain activation in males and females during viewing of erotic film excerpts.** *Human Brain Mapp* 2002;16:1–13
11. Schiffer B, Krueger T, Paul T, et al. **Brain response to visual stimuli in homosexual pedophiles.** *J Psychiatry Neurosci* 2008;33:23–33
12. Sartorious A, Ruf M, Kief C, et al. **Abnormal amygdala activation profile in pedophilia.** *Eur Arch Psychiatry Clin Neurosci* 2008; 258:271–77
13. Estellon V, Mouras H. **Sexual addiction: insights from psychoanalysis and functional neuroimaging.** *Socioaffectinve Neuroscience and Psychology* 2012;2:11814
14. Oritgue S, Bianchi-Demicheli F, Patel N, et al. **Neuroimaging of love: fMRI meta-analysis evidence towards new perspectives in sexual medicine.** *Sexual Med* 2010;7:3541–52
15. Dickinson RL. *Human Sex Anatomy:ATopographical Hand Atlas.* 2nd ed. London, UK: Balliere, Tyndall and Cox; 1949:84–109

Intelligence and memories:

BOOSTING YOUR BRAIN, PART 1: THE COUCH POTATO

The issue of failing memory worries us all, particularly as we get older. The treacherous nature of memory and its distortions caused by advancing age have been dealt with in several recent bestselling books from Julian Barnes' *The Sense of an Ending*[1] to Patrick Modiano's *L'Horizon*.[2] In both books, individuals in their late 50s or 60s attempt to reconstruct past events in different ways and with different results. Professor William Miller made light of his declining mental abilities in his book *Losing It*,[3] which he subtitled:

"In which an aging professor laments his shrinking brain which he flatters himself formerly did him noble service." Is there anything that we can do to keep from "losing it"? Can we really be "older and wiser"? In this first part, I discuss (presumed) ways to exercise your brain while avoiding exercising your body.

Dr Lawrence Katz, a former Duke University professor of neurobiology, is said to be the father of "neurobics" (yes, meaning aerobics for your neurons). Aerobics means "living in air" or, more scientifically, being able to supply enough oxygen to support increased aerobic cellular metabolism during exercise.[4] Although it is possible that we do lose some neurons with age, the "no new neuron" theory that stated that as we get older our neurons die and are not replaced has been disproved. Brain stem cells from the hippocampi and subventricular zones constantly multiply and repopulate the cortex throughout our lives. It is possible that age-related mental decline is due to a malfunction of dendrites and synapses, but the good news is that older as well as new neurons can easily sprout new dendrites. The number of oligodendrocytes also decreases with age, leading to myelin fragmentation, decreased number of myelinated fibers, and ultimately loss of axons.[5] Diffusion tensor imaging has shown that myelin achieves its greatest degree of complexity during the fifth decade of life, so it is not surprising that many of us feel that middle age is when we have been smartest. DTI has also demonstrated stronger connectivity in regions associated with better executive function.[6]

Some think that by using DTI to identify weak connections, we should be able to design specific brain exercises aimed at strengthening these regions.

Even Aristotle knew that practice was helpful in maintaining a healthy brain.[7] Traditional brain exercises include puzzles and memory jogging. Reading and even playing bingo are said to be beneficial, and maybe that is why both are popular in retirement homes. Among the many Web sites dedicated to brain exercises, *Happy Neuron*[8] lets you tailor several games according to your age, sex, and education level and then makes money by attempting to sell you "adult brain-training products" by using descriptions reminiscent of those used by a different a type of "adult" industry. This Web site claims to improve your brain by 16%

(whatever that means).

In a study published in 2002, cognitive training was effective and durable in improving cognitive function.[9] The authors reported that the benefits of training basically erased the expected mental decline during 7- and 14-year periods. This is important because prior studies had shown that the effects of mental calisthenics tend to last only 2 years. What the authors do not clearly explain is why there was no functional brain decline in the control group as would have been expected.

Thus, it is not surprising that many scientists accept the beneficial effects of brain training but warn us that these are only moderate. Additionally, you get better at what you trained for and nothing else. A recent article in *Nature*[10] reported that more than 11,400 subjects were trained several times per week in tasks designed to improve reasoning, memory, planning, and attention.

Although improvements were noted for these specific tasks, the benefits could not be transferred to other untrained tasks, even if they were cognitively closely related. This observation is akin to my children telling me that computer games make them smarter.

These games make them better at playing the games but nothing else.

Most of these brain exercises do not result in what is called long-term potentiation—the potentiation is present only while you do them regularly. The idea behind long-term potentiation is that brain exercises lead to production of neurotrophins, such as brain-derived neurotrophic factor, which stimulates dendritic sprouting and actually physically changes the brain. Neurotrophins, however, only result in neuronal branching when those same neurons are being constantly stimulated. Analogous to a muscle, if you stop your exercises, that part of the brain turns flabby. Extensive long-term repeated experiences may induce changes in brain morphology due to neuroplasticity. London taxi drivers have increased gray matter and hippocampal volumes.[11]

The longer they drive, the bigger these regions get. I wonder if these will regress in size if drivers start using a Global Positioning System. Learning music, as an adult or a child, improves verbal and working memory as well as attention and is also said to change the shape of the brain.[12]

Couch potatoes may be too lazy even to exercise mentally; so, what could be better than improving your brain by just swallowing a pill? The use of methylphenidate (Ritalin) as a neuroenhancer is widespread in American high schools and on college campuses (this is not the case in Europe and Latin America).

Students who do not have documented attention deficit disorders take it to cram for examinations. By increasing brain dopamine, they feel more alert, focused, interested, and motivated. Amphetamines in different forms stimulate the brain and are also widely used on university campuses. Adderall is one of these medications and is only legal in the United States and Canada (so again, not used for that purpose elsewhere). Students report a 2-letter improvement in their grades when taking these drugs.[13] On some campuses, up to 35% of students regularly take these pills.[14] Other "smart drugs" include phosphatidyl serine (said to activate cell to- cell communication) and vinpocetine (increases blood flow to the brain and eyes); others that should improve memory and boost intelligence are now being designed. The list of acute and chronic complications from using any of these drugs is too long to recount here, but pharmaceutical companies will continue to make and sell them rather indiscriminately because they are a huge business. In 2010, more than US $470 million of Ritalin was sold, and because we want to make all medications cheaper and more available, the company Actavis (Elizabeth, New Jersey) this year launched its generic and less expensive equivalent.[15] Commonly available in nearly all supermarkets is ginkgo biloba, which may moderately improve cognition particularly in aging individuals and patients with Alzheimer disease. This extract from plant leaves acts by increasing cerebral blood flow, and a study done at Johns Hopkins by using perfusion MR imaging showed that cerebral blood flow was mildly and globally increased in elderly individuals who took that supplement twice per day.[16] Additionally, it has been given to patients with Down syndrome and early-onset Alzheimer disease with positive results.[17]

If all of the above seems too time-consuming, too expensive, or too risky, you may want to try neurobics. Neurobics requires only that you do 2 things: experience the unexpected and use all of your senses every day.[18] Dr Katz's method has been extremely popular, and his book has been reprinted 25 times and translated into 24 languages. His coauthor, Mr. Manning Rubin, is a commercial writer who runs an advertising company and has worked for some of the biggest marketing organizations in the world, so he knows a bit about product placement; but in all fairness to the book, Dr Katz was the real thing (he died in 2005 from melanoma).

A graduate of the University of Chicago and the California Institute of Technology, he went on to become faculty at Duke University and published many articles in peer-reviewed journals.

Because neurobics can be incorporated into any of our daily activities, it sounds initially attractive, but it seems that like anything else, when you look carefully into it, it is a bit more complicated.

Here are some of the recommended activities (my comments in parentheses):

1. Shower with your eyes closed. (The book clearly warns you about balance issues, common as we get older, and water that may be hot if you are not looking at the temperature handle.)
2. Get dressed with your eyes closed. (Tried it last week and ended up putting on socks that were the wrong color.)

3. Brushing roulette, meaning brush your teeth or hair with your non dominant hand. (Be careful about hurting your gums or poking an eye.)
4. Listen to specific music while smelling something particular. (This requires a bit of planning, and I have not tried it yet because it seems too complicated and I cannot decide which smell goes well with Orff's *Carmina Burana*.)
5. Take your family to work with you. (I imagine that the mental exercise kicks in when trying to explain to them how susceptibility- weighted imaging or iterative reconstruction works.

   Beware: this can be a Health Insurance Portability and Accountability Act violation!)
6. Wear ear plugs when you sit down for a meal with your family. (I think it is the effort of lip reading that exercises your mind. Ask your spouse before doing this to avoid conflicts at home.)
7. Turn your computer monitor upside down. (When I was a resident, a very senior and very smart attending used to read the chest radiographs upside down. He was doing neurobics without even knowing it!)
8. Shop at the farmers' market instead of the Big Box supermarket.

   (You get to talk to real folks, eat healthier, and support your local economy. Not a bad idea!)

The list of somewhat wacky ideas goes on and on (e.g., create a sensory symphony in the bath, start watching *Sesame Street* again, and so forth). Dr Katz also reminds us that sex is the ultimate brain workout; but because it can also be a physical workout and this *Perspectives* does not deal with physical exercise, I will comment on it next month in Part 2. The book, *Keep Your Brain Alive*, contains detailed chapters on neurobics during your morning and afternoon commutes and neurobics while at work, at the market, at mealtimes, and at leisure. In reality, most of its recommendations involve variations on activities that we get too lazy to do as time goes by.

If you are still too lazy to do any of the above, how about just breathing? Oxygen appears to enhance memory. Higher blood oxygen saturation increases the heart rate, and both correlate with improved brain performance. In one study, subjects who inhaled oxygen 60 seconds before attempting to memorize a word list outperformed those who did not.[19] Oxygen cannot be given in advance; it has to be inhaled just before the task, suggesting it is increased blood oxygen saturation that plays a role in memory consolidation. Every time I lecture, I look at my presentation and try to memorize all of its slides so that I can "bridge" them while speaking. Because I do not carry an oxygen tank with me, just recently I tried hyperventilating before a lecture, and the results were that I felt dizzy and shaky, my hands were tingling, and I ended up giving a terrible lecture. There are better ways to improve your memory, albeit involving some work and discipline, and a different essay will deal with these.

REFERENCES

1. Barnes J. *The Sense of an Ending*. New York: Knopf; 2011
2. Modiano P. *L'Horizon*. Paris, France: Editions Gallimard; 2010
3. Miller W. *Losing It*. New Haven, Connecticut: Yale University Press; 2011
4. Aeorobic exercise. Wikipedia. http://en.wikipedia.org/wiki/Aerobic_exercise. Accessed April 16, 2012
5. Marner L, Nyengaard JR, Tang Y, et al. **Marked loss of myelinated nerve fibers in the human brain with age.** *J Comp Neurol* 2003;462:144–52

6. Wen W, Zhu W, He Y, et al. **Discrete neuroanatomical networks are associated with specific cognitive abilities in old age.** *J Neuroscience* 2011;31:1204–12
7. Aristotle. De memoria et reminiscentia. http://etext.lib.virginia.edu/ toc/modeng/public/AriMemo.html. Accessed April 16, 2012
8. Happy Neuron: 100,000,000 games played. Brain fitness for life. http://www.happy-neuron.com. Accessed April 16, 2012
9. Ball K, Berch DB, Helmers KF, et al. **Effects of cognitive training with older adults: a randomized controlled trial.** *JAMA* 2002;28:2271–81
10. Owen AM, Hampshire A, Grahn J, et al. **Putting brain training to the test.** *Nature* 2010;465:775–79
11. Maguire EA, Woollett K, Spiers HJ. **London taxi drivers and bus drivers: a structural MRI and neuropsychological analysis.** *Hippocampus* 2006;16:1091–101
12. Memory improvement. Wikipedia. http://en.wikipedia.org/wiki/Memory_improvement. Accessed April 16, 2012
13. Popping pills a popular way to boost brain power. http:// **694** Editorials Apr 2013 www.ajnr.org www.cbsnews.com/2100-18560_162- 6422159.html?pageNum_ 2&tag_contentMain;contentBody. Accessed April 16, 2012
14. Talbot M. Can a daily pill really boost your brain power? 60 Minutes. http://www.guardian.co.uk/science/2009/sep/20/neuroenhancers-usbrain- power-drugs. Accessed April 16, 2012
15. Ritalin. http://www.evaluatepharma.com/Universal/View.aspx? type_Entity&entityType_Product&id_10818&lType_mod Data&componentID_1002. Accessed April 16, 2012
16. Masahyekh A, Pham DL, Yousem DM, et al. **Effects of ginkgo biloba on cerebral blood flow assessed by quantitative MR perfusion: a pilot study.** *Neuroradiology* 2011;53:185–91
17. Buckely F, Sacks B. Drug treatment improves memory in mice. http://www.down-syndrome.org/updates/2037. Accessed April 16, 2012

18. Katz LC, Rubin M. *Keep Your Brain Alive*. New York: Workman Publishing Company; 1999
19. Chung SC, Lim DW. **Changes in memory performance, heart rate and blood oxygen saturation due to 30% oxygen administration.** *Int J Neurosci* 2008;114:593–606

BOOSTING YOUR BRAIN, PART 2: THE HARD WAY

In my previous essay, I enumerated several ways to exercise your brain and boost its capacity without expending too many calories or too much effort. You may recall that the results of such methods were controversial and minimally helpful at best.

Before starting this vignette, I need to confess that I used to hate physical exercise. I remember failing the physical education courses in high school several times and spending long vacations in the library writing reports on the history of the Olympic Games to make up for my bad grades. When I turned 50 years of age, I made an appointment to see a doctor and was surprised at finding myself in the Geriatric Medicine Building. He recommended, as he does for all "senior" individuals, that I exercise, and now I do it regularly at least 4 times per week. Truth be told, I feel more attentive and even smarter if I do it before going to work.

Exercise makes us feel better physically and mentally, especially as we get older. In 1 study, the benefits of exercise were not affected by socioeconomic factors, presence of disease, body mass index, smoking, cohabitation, or disability and had clear benefits in elderly subjects compared with those who did not exercise.[1]

Cohabitation may be a stressful situation, and it has been shown that animals that are exercised regularly are better at living with other animals than those who are not.[2] Although there are no human data to support similar benefits, it seems logical that we, too, would benefit in this regard. It is encouraging that one does not need to exercise strenuously to improve the brain. In older individuals, even light activities such as walking around the block, gardening, and cleaning are mentally beneficial. So perhaps there is some truth when people say that walking helps "clear their heads." Walking and staying active are said to change the dynamics of aging and to delay the onset of dementia by about 5–10 years.[3] Weight-lifting, even light weights, has been shown by functional MR imaging to increase brain cortical activation.[4]

A recent article in *Nature* explains that exercise induces the process of autophagy.[5] Autophagy is a lysosomal function that permits cells to recycle waste products and contributes to organelle and protein quality control. Autophagy protects not only against neurodegeneration but also against cancer, infections, aging, and insulin resistance. In lay language, cells become full of trash; and unless autophagy is turned on, they clog up and die.

The contrary may also be true: If autophagy systems are genetically abnormal, no amount of exercise will improve one's health.

In addition to exercise, starvation turns on this process and it is well-known that starved animals live longer. Lack of exercise may lead to obesity and result in development of adult-onset diabetes.

In animals, only a significant amount of exercise will result in autophagy conferring a protective effect from developing diabetes.

[6] Diabetes is well-known to slow down the brain and lead to mental decline. This is particularly true if its onset occurs before 65 years of age, the disease duration is more than 10 years, it is the insulin-dependent type, and cerebral vascular complications (leading to stroke) are present. Exercise and good diet are the 2 best ways to prevent obesity and diabetes.

Obesity affects the brain in many ways, even during infancy and teenage years. Obesity leads to faster bone and sexual maturation in both sexes, and its effects on brain maturation were recently studied.[7] During healthy adolescence, there is considerable synaptic pruning and increased myelination, particularly in the frontal and parietal lobes. Obese teenagers show abnormalities in white matter, particularly affecting the orbitofrontal regions.

This translates into lower brain volumes in those areas, behavioral disinhibition, and lower performance on cognitive tests independent of the presence of diabetes. It is intriguing that at least some types of fat may be disposed of in the absence of exercise. A study shows that in response to cold, brain activity increases and this leads to activation and usage of brown fat to maintain corporal thermogenesis.[8] Because brown fat is available only in small volumes in only some parts of the body, even using it all up does not result in a significant weight loss nor does it protect from diabetes.

Muscles and brain power are generally not believed to go hand-in-hand, but the truth is that they do. Aerobics and running do great things for the brain. Weight-lifting increases the amount of brain-derived neurotrophic factor (BDNF), which sparks neurogenesis.

The need to make split-second decisions during sports extends to everyday activities in most athletes.[9] Also, exercise increases cerebral blood flow. In a study done by my colleagues here at the University of North Carolina, aerobic activity in elderly subjects was associated with lower arterial tortuosity and an increased number of small vessels, leading them to conclude that exercise may contribute to healthy brain aging.[10]

Human adult neurogenesis occurs in the dentate nucleus of the hippocampus, the subventricular zone, and the olfactory bulb and possibly also in the neocortex, striatum, amygdala, and substantia nigra.[11] It is thought that as we age we produce fewer new cells in the hippocampi and that this contributes to cognitive decline.

Experiments in elderly animals have shown that running increases hippocampal neurogenesis, which, in turn, is capable of reversing even fixed neurologic deficits that were present before exercising began.[12] These authors concluded that voluntary exercise ameliorates some of the deleterious effects of aging. Not only does exercise foster hippocampal neurogenesis but increased local production of BDNF supports growth, development, maintenance, and survival of neurons there. Similar to a previously mentioned article, better arteries were found in the hippocampi of elderly humans who exercise, and animals forced to run demonstrated remarkable hippocampal angiogenesis in response to physical exercise.[13] MR imaging– generated maps of cerebral blood volume show increased perfusion to the hippocampi of humans who exercise, a finding opposite to that seen in patients with Alzheimer disease.[14] Many fitness centers now offer aerobics for the elderly in programs called "elderobics," with the idea of promoting all of the above-mentioned benefits.

While exercising can improve your brain's performance, just thinking about muscle training can result in increased physical strength. In a fascinating study, subjects performing mental training of their hip flexors increased their strength by 24% compared with the 28% gained by actual exercise.[15] These experiments have been repeated by different authors always reaching similar conclusions.

Thus, just imagine what happens if you exercise and think about it, too! In the United States, nearly 50% of children between 12 and 21 years of age do not exercise, and less than 25% report getting 30 minutes of physical activity per day.[16] Exercise by itself may not be enough to prevent mental deterioration because other factors such as smoking, hypertension, diabetes, obesity, low levels of education, and diet also play an important role.

In Part 1 of "Boosting Your Brain," I mentioned sex. In this regard, sex offers 2 benefits: exercise and a naturally rewarding activity, both of which have direct positive effects on the brain (at least studies in rats prove these benefits, and it is reasonable to believe that they extend to us humans). Conversely, negative stress suppresses adult neurogenesis. Rats that had daily sex for only 14 days showed new hippocampal neurons.[17] Frequent sex induced continuous neuronal growth as well as growth of dendrites.

If exercising routinely is hard, just try dieting. The types of diet that exist are simply too many to consider here and are confusing (and outright bizarre, too; for example: the alkaline diet, lemonade diet, banana diet, and the purple food one). We know that a balanced diet is essential to maintain our brain's health. Antioxidants appear to directly benefit memory and are found mainly in flavonoids, which interact with BDNF and improve long-term potentiation. Like ginkgo biloba (see Part 1), flavonoids increase cerebral perfusion.

Grapes, tea, cocoa, and blueberries are good for the mind. Ingesting glucose is also great, but only in moderate amounts; excessive glucose ingestion impairs brain function.[18] Diets rich in fat also impair memory and affect the integrity of dendrites in the hippocampi. It has been postulated that fats promote an early onset of Alzheimer disease, whereas some types of fats, such as those found in fish oils, conversely may help prevent dementia.[19] As mentioned before, starvation is good for the brain, whereas high-calorie diets promote neuronal dysfunction and degeneration.[20]

Recent data indicate that physical inactivity is linked to 21% of Alzheimer cases in the United States and 13% worldwide.[21] These numbers speak badly of our lifestyle here in the United States; but in Europe, only 40% of individuals exercise once a week and countries where citizens exercise the least often are Bulgaria, Greece, and Italy, with only 3% of their citizens exercising regularly.

In contrast, 75% of Swedish and Finnish individuals exercise at least once per week.[22] Asian countries also need their populations to exercise more; the Taiwanese are probably the least physically active in the entire world.[23]

Because many people exercise and then go to work and sit down all day long, most benefits from their initial physical activity will be cancelled out. Recent information indicates that regardless of exercising, if one sits more than 8 hours per day (typical for American adults), there is a 15% greater risk of dying in the following 3 years.[24] This important observation makes one realize that after exercising, the rest of day needs to be filled with physical activities and not sedentary ones to preserve the benefits of the initial workout.

If you want to start exercising, experts tell you the following: keep it simple, 1 goal at a time and preferably a measurable one; develop motivation (as if keeping your mind sharp was not enough!); and report your progress to someone. Just think about this: If I was able to start doing it, so should you!

REFERENCES

1. Bertheussen GF, Romundstad PR, Landmark T, et al. **Associations between physical activity and physical and mental health: a HUNT 3 study.** *Med Sci Sports Exerc* 2011;43:1220–28
2. Reynolds G. **Well. Why exercise makes us feel good.** http://well.blogs.nytimes.com/2011/07/06/why-exercise-makes-us-feel-good/

3. Reynolds G. **Well. How exercise can keep the brain fit.** http://well.blogs.nytimes.com/2011/07/27/how-exercise-can-keep-the-brain-fit/
4. Liu-Ambrose T, Nagamatsu LS, Voss MW, et al. **Resistance training and function plasticity of the aging brain: a 12-month randomized controlled study.** *Neurobiol Aging* 2012;33:1690–98
5. He C, Bassik MC, Moresi V, et al. **Exercise-induced BCL2-regulated autophagy is required for muscle glucose homeostasis.** *Nature* 2012;481:511–15
6. Reynolds G. **Well. Exercise as housecleaning for the body.** http://well.blogs.nytimes.com/2012/02/01/exercise-as-housecleaning-for-the-body/? scp_1&sq_exercise%20as%20housecleaning%20for %20the%20 body&st_cse. Accessed May 30, 2012
7. Cazettes F, Tsui WH, Johnson G, et al. **Systematic differences between lean and obese adolescents in brain spin-lattice relaxation time: a quantitative study.** *AJNR Am J Neurodiol* 2011;32:2037–42
8. Miao Q, Zhao XL, Zhang QY, et al. **Stability of brain glucose metabolism following brown adipose tissue inactivation in Chinese adults.** *AJNR Am J Neuroradiol* 2012;33:1464–69
9. Chaddock L, Neider MB, Voss MW, et al. **Do athletes excel at everyday tasks?** *Med Sci Sports Exerc* 2011;43:1920–26
10. Bullitt E, Rahman FN, Smith JK, et al. **The effect of exercise on the cerebral vasculature of healthy aged subjects as visualized by MR angiography.** *AJNR Am J Neuoradiol* 2009;30:1857–63
11. Sierra A, Encinas JM, Maletic-Savatic M. **Adulthumanneurogenesis: from microscopy to magnetic resonance imaging.** *Front Neurosci* 2011;5:47
12. van Praag H, Shubert T, Zhao C, et al. **Exercise enhances learning and hippocampal neurogenesis in aged mice.** *J Neurosci* 2005;25:8680–85
13. Van der Borght K, Kobor-Nyakas DE, Klauke K, et al. **Physical exercise leads to rapid adaptations in hippocampal vasculature: temporal**

dynamics and relationship to cell proliferation and neurogenesis. *Hippocampus* 2009;19:928–36

14. Pereira AC, Huddleston DE, Brickman AM, et al. **An in vivo correlate of exercise-induced neurogenesis in the adult dentate gyrus.** *Proc Natl Acad Sci U S A* 2007;104:5638–43
15. Singn G. **Merely thinking about exercising can increase physical strength.** http://www.examiner.com/article/merely-thinking-aboutexer-cising- can-increase-physical-strength. Accessed June 26, 2012
16. **The human brain: renew—exercise.** http://www.fi.edu/learn/brain/exercise.html. Accessed May 30, 2012
17. Leuner B, Glasper ER, Gould E. **Sexual experience promotes adult neurogenesis in the hippocampus despite an initial elevation in stress hormones.** *PLoS One* 2010;5:e11597
18. Hall JL, Gonder-Frederick LA, Chewning WW, et al. **Glucose enhancement of performance on memory tests in young and aged humans.** *Neuropsychologia* 1989;27:1129–38
19. Puglielli L, Tanzi RE, Kovacs DM. **Alzheimer's disease: the cholesterol connection**. *Nature Neurosci* 2003;6:345–51
20. **Memory improvement.** Wikipedia. http://en.wikipedia.org/wiki/Memory_improvement#cite_note-42. Accessed May 30, 2012
21. Belluck P. **Grasping for any way to prevent Alzheimer's.** *The New York Times.* July, 25, 2011. http://www.nytimes.com/2011/07/26/health/26alzheimer.html
22. European Union. www.europa.eu. Accessed May 30, 2012
23. **Official claimsTaiwanesepeople least physically active.** *TheChinaPost.* September 19, 2011. http://www.chinapost.com.tw/taiwan/foreign-affairs/ 2011/09/19/317069/Official-claims.htm. Accessed May 30, 2012
24. Reynolds G. **Well. Meet the active couch potato.** http:// well.blogs.nytimes.com/2012/04/04/meet-the-active-couch-potato/?pagemode_print. Accessed May 30, 2012

THINKING IN DIFFERENT DIRECTIONS

A few days ago, we were having our weekly case conference and we saw a patient with an interesting intracranial lesion. I asked a resident to look at the clinical record and find out what the discharge diagnosis was. He then proceeded to inform me that it was a meningioma. "How do they know if they have not biopsied it?" I asked. After much searching, our trainee found out that the neurosurgeons had used our initial impression as the final diagnosis, and slowly we had all begun to assume that this was indeed the confirmed diagnosis and kept quoting it on our own reports.

This is an example of thinking that begins and ends with an assumption (often wrong), also known as circular or paradoxical thinking and in logic called a "logical fallacy."[1] It is my impression that in imaging and in medicine in general, we spend a considerable amount of time engaged in this type of reasoning and that this process is more common now than in the past, perhaps because of the repetition ("cutting and pasting") that is found in patient medical records. In circular thinking, a conclusion cannot be proved false or true if it arose from a false premise. Because repeating a statement in circular fashion seems to make it stronger, circular thinking ends by creating statements that sound true and gain wide support (thus, the above-mentioned patient now carries a diagnosis of "meningioma"). There is no doubt that circular thinking is dangerous and that we must do our best to avoid it.

The opposite of circular thinking is linear (vertical) thinking.

In this type of reasoning, progress is made in a step-by-step fashion and a response to each step must exist before advancing to the next one. Although linear thinking advances by logic, it is by its own nature highly focused on single pathways and as such tends to ignore other possibilities and alternatives. Linear thinking is basically a binary process in which answers are "Yes" or "No"

(correct or incorrect), excluding all considerations beyond these 2 responses. These features make it fast, organized, and sequential and therefore it is the most common type of thought process used.[2] People generally regard linear thinking as an honest, mature, and intelligent process when in reality it lacks ingenuity, innovation, and originality. Similar to circular thinking, linear thinking is characterized by repetition and is, in the long term, detrimental to intellectual advancement.

Where linear thinking is a "safe" process, a third type of reasoning called lateral (horizontal) thinking is risky, uneven, adventurous, more difficult, and not widely accepted. Lateral thinking views a problem from multiple perspectives, many of them random.

Because lateral thinking is based on discovery and exploration of spontaneous events, it is the opposite of linear thinking: slow, disorganized, and non-sequential. Lateral thinking teeters close to the edge of disaster because it is greatly affected by luck and chance and may easily turn into chaos. Most individuals are not organized enough to use it and rapidly become overwhelmed by the choices it offers. The brightest individuals know when to use vertical and lateral thinking and avoid circular reasoning. In popular culture, linear thinking is linked to men while lateral thinking is linked to women.

Howard E. Gardner, a world-famous professor of cognition and education at Harvard University, formulated the concept of multiple intelligences.[3] We humans have different ways of learning and processing information and thus we are different and independent from each other. Although those who favor this concept oppose the idea of a "general intelligence factor," it is likely that all individuals share both, that is, they are smart as individuals but also share a collective intelligence that makes them similar to all other human beings. Dr. Gardner has separated intelligence into the following categories: linguistic, logical-mathematical, musical, spatial, bodily/kinesthesic, interpersonal, intrapersonal, naturalistic, and possibly existential (after much thinking I have come to the conclusion that there must be others because I believe I do not possess any of these!). However, I agree with him when he states that education (not only in America but worldwide) is based mostly on logical (mathematics) and linguistic (language arts) intelligence and that current methods for assessing intelligence (such as IQ tests) measure only these 2 features. This brings up the inadequacy of the current schooling systems that disregard other types of intelligence. Most current education (and research) depends on mainly linear thinking.

A fascinating endeavor that encourages folks to express their different intelligences and to think laterally is TED.com (TED stands for: technology, entertainment, design). This nonprofit organization that was started in 1984 contains more than 1400 (as of this writing) varied and exciting conferences by some of the world's smartest and most diverse and laterally thinking individuals.

For a fantastic account of how it works, I suggest reading Nathan Heller's article in *The New Yorker* titled "List and Learn."[4]

The most viewed TED conference (more than 15 million times) is one given by British education specialist Sir Ken Robinson in 2006 (a newer one was posted in May 2010 and has been viewed nearly 4 million times).[5] Robinson argues that university professors educate students to become, well... university professors in a process so linear that it kills all creativity and discourages many students from exploring alternative avenues. He also calls attention to the ever-diminishing value of education degrees (and those of us who live in university towns know that sometimes all that a

PhD gets you is a better waitressing job). The rigidity of school systems that are based on mathematics and linguistics results in linear thinking stifling the creativity associated with lateral thinking and is thus harmful to society.

In an article in *The New York Times,*[6] Andrew Hacker explains why more than one-third of high school students fail algebra and states that difficulty with mathematics may be responsible for up to 45% of high school dropouts in the United States. Aptitude tests such as the American SAT (Scholastic Aptitude Test) and the ACT (American College Testing) concentrate in measuring 2 subjects: mathematics and linguistics (the pillars of linear thinking, as stated previously). Mr. Hacker proposes that perhaps just basic algebra and what he astutely calls "citizen statistics" may be enough for most us, whereas more advanced courses such as calculus should be reserved for fewer, gifted individuals who seek careers that depend on the understanding of higher mathematics.

As Sir Ken Robinson states, "We are educating people out of their creative capacities."

Does studying liberal arts and the humanities make us better physicians? I believe it does. I have been unable to notice any differences with respect to knowledge of biologic sciences in our daily work between residents who come from a "hard" science background and those with a liberal arts education, and I find that personally I like the latter better. Medical schools are aware of this, and some such as Boston University and Brown University encourage this type of liberal arts curriculum and reserve a number of places in their medical schools for these individuals. More than 40% of medical students at the University of Pennsylvania come from non-premed backgrounds.[7] The liberal arts may also be useful to medical students, and the Mount Sinai School of Medicine in New York has a specific humanities and medicine program.

The separation of liberal arts from sciences is, in my opinion, damaging and ends up suppressing the human qualities of many excellent and caring young individuals. Because liberal arts are characterized by lateral thinking, bringing these individuals into our world of linear thinking will prove to be beneficial for all.

I am not aware of imaging methods having been used to study these different types of thinking. There are, however, several principles that control all human thought processes.[8]Abasic principle of thinking is that it is the product of concurrent brain activity in multiple regions that together form a large-scale cortical network.

This is a type of functional connectivity that has been documented in thousands of fMRI reports. Also, each cortical region can perform multiple functions, and these same functions can also be performed by different regions, an observation that may explain thought (and function) plasticity. Rather than a strict linear or vertical organization, the brain prefers a lateral or horizontal organization that serves as its own backup and redundant system.

Unfortunately, each cortical region can only do so much and thus has a limited capacity. Conversely, these constraints force other parts of the brain to collaborate, and this helps it adapt to many situations. The topologies of large-scale networks are in constant flux, adapting themselves to the demands of tasks. The brain is not dumb: it uses the minimum amount of resources needed for each activity, but, if one network becomes insufficient, additional ones are immediately recruited. The brain's topology has 2 components: membership and connectivity, and both are in constant flux. Just as the Internet does, the brain also has a limited bandwidth, resulting in a finite amount of resources that it can use.

This bandwidth, up to a certain extent, varies from individual to individual and thus some are more successful in multitasking than others. Increased brain bandwidth seems to be connected to lateral thinking.

Lateral thinking is important and is not used sufficiently in the sciences, but this is beginning to change. Of course, we radiologists can take it to a silly extreme, as seen in a recent advertisement that intended to recruit a lateral-thinking technologist for a vertical MR imaging (upright) scanner![9] Radiologists actually think dimensionally, and 2D and 3D processes play an important role in the interpretation of images in which it all begins as the former and ends as the latter. I like to think of this as a process that also begins vertically and then branches horizontally. Some of our trainees have more trouble making this transition and thus take longer to learn the specialty. It is possible that some may survive and graduate not being able to think 3-dimensionally, but they will never survive if they think circularly.

REFERENCES

1. **Circular reasoning.** http://en.wikipedia.org/wiki/Circular_ reasoning. Accessed on May 2, 2013
2. **Is linear thinking bad?** http://www.andyeklund.com/creativestreak/2012/06/is-linear-thinking-bad.html. Accessed on May 2, 2013
3. http://www.howardgardner.com. Accessed on May 2, 2013
4. Heller N. **List and Learn.** *The New Yorker,* July 9, 2012
5. http://www.ted.com/talks/ken_robinson_says_schools_kill_creativity.html. Accessed on May 2, 2013
6. Hacker A. **Is algebra necessary?** *The New York Times,* July 28, 2012. http://www.nytimes.com/2012/07/29/opinion/sunday/is-algebraneces-sary. html. Accessed on May 2, 2013

7. Kliff S. **Why medical schools like to accept STS majors.** *Newsweek*, September 10, 2007
8. Just MA, Varma S. **The organization of thinking: what functional brain imaging reveals about the neuroarchitecture of complex cognition.** *Cogn Affect Behav Neurosci* 2007;7:153–91
9. http://www.radmagazine.co.uk/JobPDFs/441115.pdf. Accessed on May 2, 2013

Personal improvement:

THE 3 PILLARS OF HEALTH

For many years, I undervalued sleep. Why sleep when one can be up and about? Well, several things have recently caused me to change my mind. Our understanding of sleep and how it affects our brains and bodies is rapidly changing, and personal fitness devices help us keep track of these effects.

Approximately 2 months ago, I bought a personal fitness device; one of those plastic bands worn around a wrist that have lately been in the news. Most of these devices rely on accelerometers and altimeters (such as those used in the iPhone and iPad that allow one to play games by twisting and moving the devices) and store their data in small computers that then take this information to calculate different parameters. With regard to the types of physical activity analyzed, the performances of these devices vary and I tend to concentrate on trends rather than specific daily measures (but by now I know exactly how much I walk when on clinical service from my office to the reading room and how much I move during academic days).

With regard to sleep, my device does something called "sleep efficiency tracking." After communicating via Bluetooth and Wi-Fi with my telephone, I get to see the following: total sleep time, hours of deep and light sleep, and times and duration of periods in which I was awake. Every few days, it gives me messages informing that I am sleeping well or need to sleep more such as, "If you sleep more than 7 hours today, tomorrow you will feel more rested, confident, beautiful, and smart!" How can one resist such messages?

In addition (I have not tried it yet), the device has a smart waking alarm, meaning that it will wake me within 30 minutes of a selected time based on my REM cycles. Apparently, this eliminates the disconcerted feeling of being woken up in the middle of a dream. In reality, I think that most sleep information it provides is fairly vague but, again, trends are what count.

One message that my device has sent me several times states that sleep (the National Sleep Foundation recommends 7–9 hours per day for adults) helps one lose weight, and this does make sense. If one is not sleeping well, one is spending more time awake, which increases the chance of eating and induces fatigue that leads to more eating (especially carbohydrates) and lesser physical activity.

Sleep deprivation produces a "double hit" to the brain in this regard: there is a sharp reduction in frontal lobe activity leading one to misjudge hunger and eat more junk and also leading to an exaggerated activation of reward centers (mostly the amygdalae) after eating. One meta-analysis of nearly 700 published studies showed that both adults and children who are short sleepers have an increased risk of obesity.[1] In a different study, 12 men were allowed a full night of sleep (8 hours) followed by a partial night of sleep (4 hours); after the latter, the men were hungrier upon waking up and ate more during the day (22%).[2] Acute partial sleep leads to increased serum levels of ghrelin (a hunger hormone) and decreased levels of leptin (a satiety hormone). Paradoxically, sleeping less leads to increased metabolism and extra burning of calories (but on average just a paltry 111 calories per day). However, the compensatory intake of food exceeds this amount and results in a net weight gain.[3] Partial sleepers tend to consume approximately 600 more calories than full-night sleepers, with most of those calories coming from chocolate, desserts, and potato chips.[4] Poor food choices are common in the sleep deprived. One of the authors of this last study speculated that adenosine, which is a brain metabolic by-product, possibly disrupts neural function by accumulating in the sleep-deprived brain.[5] It is conceivable that adenosine accumulation affects the way we perceive food. Because sleep clears the brain of adenosine, sleep is the equivalent of chemically rebooting our brains.

Sleep must be terribly important because all animals on earth do it. A recent article published in *Science* starts to shed light on its importance. Rats undergoing natural sleep or sleep induced by anesthesia increased their brain's interstitial spaces by 60%.[6] This leads to significant exchanges between CSF and interstitial fluid that result in increased rates of beta-amyloid clearance during sleep. Apparently, amyloid accumulates during the day and sleep disposes of it at night. This system is thought to be analogous to the lymphatic system that clears metabolic waste from the rest of the body, hence its name: glymphatic system (glial_ lymphatic).

Sleep promotes memory consolidation and that is why this process is abnormal in the elderly (who sleep less) and in those with diseases such as Alzheimer, in which amyloid accumulates.

The term "glymphatic system" derives from the fact that the glial cells are the ones mostly responsible for the exchange of fluids through their membranes. The glymphatic system is a high energy system and that is why it cannot work when the brain's energy is being used for other activities such as being awake.

Wakefulness decreases the brain's fluid exchange capacity by approximately 95%. Norepinephrine probably serves to control this fluid exchange. The authors of the *Science* article speculated that increased norepinephrine when awake results in increased cell volume and conversely decreased interstitial volume. Animals given adrenergic receptor antagonists show increased CSF influx into their brain's interstitial space. Thus, rather than sleep, it is probably wakefulness that primarily regulates the glymphatic system.

The problem is that once amyloid starts to accumulate in the brain, sleep cycles are further upset, and the insomnia that occurs with Alzheimer and other dementias further exacerbates the lack of clearance of the brain's noxious substances.

Although the concept of the glymphatic system is today believed to be related only to amyloid, it is possible that it also plays a role in the disposal of other proteins such as tau and alphasynuclein that are involved in other neurodegenerative disorders, especially Parkinson disease. Other possible implications of this mechanism are the increase in migraines and seizures in patients who sleep poorly. It is conceivable that, in the future, instead of treating these diseases directly, we could design medications that enhance the functioning of the glymphatic system and radiotracers that could be used in vivo to assess their effects. How CSF gets into the brain is still unclear, possibly through its external and ventricular surfaces and the perivascular spaces.

PET imaging with the Pittsburgh compound (PiB-PET) has shown that shorter sleep duration is associated with higher amyloid brain burden.[7] Other studies have shown that unfragmented sleep reduces the risk of Alzheimer disease (and decreases the development of neurofibrillary tangles) and diminishes normal age-related cognitive decline.[8] Apnea is another factor that prevents adequate sleep consolidation and though it was thought to affect mentation by vascular effects on the brain, the worse cognition associated with it could be caused by interruptions of the glymphatic system function. Of course, one's inability to consolidate sleep is multifactorial and includes comorbidities and genetic and environmental factors among others. Perhaps information gathered by personal fitness devices on the sleep patterns of millions of users will shed more light onto the relationship between sleep and successful aging. Over 60% of all adults report trouble sleeping at night, thus the implications of the relationship between sleep and cognition are staggering.

Regardless of any scientific evidence, sleep is a hot item and even a new, short e-novella by famous (and very good) author Karen Russell deals with the subject.[9] In this short book, America experiences an epidemic of insomnia and a large corporation decides that sleep is a commodity. Healthy sleepers are urged to "donate" their sleep to those less fortunate. The fact that sleep is indeed a commodity is now being used by industries successfully.

Personal fitness and activity tracking devices generated over US $290 million last year in sales and this is expected to double soon.

Today, there are so many brands that manufacture these personal fitness devices that choosing one is difficult, particularly if one did not sleep well the night before.

So, what did I learn about my sleep in the last 2 months of wearing my Up24 band? Well, only good news. I fall asleep faster and sleep longer than I thought and have longer periods of deep sleep than others in my "team" (yes, you cannot only compete for the most activities but also for the most and best sleep). I have also become more aware as to how much people care about how I well sleep: hotels offer me better mattresses, more pillow choices, high efficiency sheets and pillow covers, mood lightning and soothing sounds, calming pulse-point oils, and some will even call before I go to sleep to remind me to turn my electronic devices off (the blue wavelength light these devices emit affects the secretion of melatonin more powerfully than any other type of light). Airlines lagged behind hotels and now that, at least in business class, the food is improving, they are concentrating on sleep and offering natural-fiber bedding, flat-bed seats, noise-cancellation headphones, and "radio" stations with only white noise. Why the industry is doing all of this is not clear to me and in a recent article in the *New York Times*, the executive director of the Harvard Medical School Division of Sleep Medicine said, "Sleep is the enemy of capitalism, you can't produce or consume when you're asleep."[10]

In the same article, Dr. Sanna also says that we need to stop thinking of sleep as a commodity and a lifestyle choice, but rather as the third pillar of health together with diet and exercise. Could it be that living and sleeping better and longer is becoming more important than just accumulating stuff? I do not think so, but tonight I will go to sleep earlier hoping that it will help me be a smarter and healthier neuroradiologist tomorrow.

REFERENCES

1. Cappuccio FP, Taggart FM, Kandala NB, et al. **Meta-analysis of short sleep duration and obesity in children and adults.** *Sleep* 2008; 31: 619–26
2. Brondel L, Romer MA, Nouques PM, et al. **Acute partial sleep deprivation increases food intake in healthy men.** *Am J Clin Nutr* 2010;91: 1550–59
3. Markwald RR, Melanson EL, Smith MR, et al. **Impact of insufficient sleep on total daily energy expenditure, food intake, and weight gain.** *Proc Natl Acad Sci U S A* 2013;110:5695–700
4. Greer SM, Goldstein AN, Walker MP. **The impact of sleep deprivation on food desire in the human brain.** *Nat Commun* 2013;4:2259

5. O'Connor A. How sleep loss adds to weight gain. *New York Times*. August 6, 2013. http://well.blogs.nytimes.com/2013/08/06/how-sleeploss- adds-to-weight-gain. Accessed March 27, 2014
6. Xie L, Kang H, Xu Q, et al. **Sleep drives metabolite clearance from the adult brain.** *Science* 2013;342:373–77
7. Malkki M. **Alzheimer disease: sleep alleviates AD-related neuro-pathological processes.** *Nat Rev Neurol* 2013;9:657
8. Lim, AS, Yu L, Kowgier M, et al. **Modification of the relationship of the apolipoprotein E _4 allele to the risk of Alzheimer disease and neuro-fibrillary tangle density by sleep.** *JAMA Neurol* 2013;70:1544–51
9. Russell K. *Sleep Donation*. New York: Atavist; 2014
10. Rosenbloom S. From airlines to hotels, a quest to help you sleep. *New York Times*. March 26, 2014. http://www.nytimes.com/2014/03/30/ travel/ from-airlines-to-hotels-

OF GIRTHS AND BRAINS

It is now official: We Americans are no longer the heaviest in the Western World. This ignominious claim belongs south of the border, to Mexico. The obesity rate of Mexicans (32.8%) has now surpassed that of Americans (31.8%). However, that is just obesity; if you take all of those who are overweight, the rate goes up to 70%. Mexican child obesity has tripled in the last 10 years, and though a difficult concept to grasp, many obese individuals are also malnourished. Eating junk food makes one fatter but does not provide the necessary nutrients to be healthy. This weight gain trend is reflected throughout Latin America and other parts of the developing world. Lack of education and a desire to imitate the United States have led to this global problem, and now, being overweight far outstrips being underweight throughout the entire world (a landmark change in humanity that occurred in 2000).[1]

In Nauru and the Cook and Marshall Islands, obesity rates have reached 71%, 64%, and 46%, respectively, the highest rates worldwide.

[2] High ingestion of fried foods is probably the main culinary culprit, but Mexicans are also the highest consumers of sugary soft drinks in our hemisphere: 43 gallons (163 L) per person per year.

Very soon, the world will have more than 900 million obese individuals.

Obesity predominantly affects our 2 largest "minorities":

Hispanics and African Americans; and despite what some reports say, their obesity rates have not changed significantly in the last few years.[3] Because Hispanics and African Americans will soon together account for most of our population, it is very disturbing that by 2048, the entire US population will be obese if we do not stop this trend!

Apart from probably being unable to fit these patients into our MR imaging and CT units, neuroradiologists will become very involved in their care. Two of the most common obesity-related disorders are stroke and diabetes, with all of their neurologic complications.

[4] The current rate of stroke in the general population is about 0.5% per year, while in diabetics, it reaches 1.2% per year.

In patients with diabetes, the risk of stroke increases with age, heart disease, previous stroke, smoking, and—not surprisingly—waist circumference. Of course, even without diabetes, stroke is more common in Hispanics and African Americans. Because the older population in the United States will increase from 13% to 19% and Hispanics, from 14% to 19% by 2025, stroke will become even more prevalent (the African American population will experience no significant growth and is expected to remain at the current 13% until about 2050).[5]

The cost of caring for patients with obesity is staggering. Currently, more than US $190 billion is spent on it every year (21% of all annual medical spending in the United States), and even if obesity rates remained stable, that cost will increase to $550 billion by 2030. About $14 billion is spent every year caring for obese children. Disease due to obesity costs businesses about $5 billion in absenteeism every year. While Mexico is also a big country, its gross national product is only 13% of ours, so it faces a considerable challenge dealing with its obesity problem.

Although the main impetus for eating is a negative energy balance, many other non-homeostatic factors, such as attractiveness of food, time and season (we tend to eat more during winter), and emotion (we eat more when sad or depressed), all affect our intake. An interesting observation is that these non-homeostatic factors are so strong that they generally override the homeostatic ones.[6] Different brain regions are involved for each of these mechanisms.

The hypothalamus, parabrachial nucleus,* and the nucleus of the tractus solitarius are all in charge of the homeostatic mechanisms. Various cortical locations, the amygdalae, ventral striata, hippocampi, and the substantia nigra, are in charge of the non-homeostatic mechanisms and have complex connections with those in charge of homeostatic control. fMRI can identify activity in all of these regions. In one experiment, levels of peptide YY (PYY) in the bowel were manipulated and its effect was observed by fMRI. PYY activates vagal nerve afferents that activate homeostatic mechanisms and result in a reduction of food intake. fMRI is capable of showing that the homeostatic circuit is activated when the levels of PYY are manipulated. Another substance, leptin, a food-intake-reducing hormone, also results in brain changes measurable with fMRI. In a leptin-deficient state, individuals show activation of parts of the non-homeostatic circuit and their desire for food increases, but when given leptin, the desire for food goes away, meaning the homeostatic mechanisms return.

There are 2 types of thin individuals: those who do not gain weight (obesity-resistant) and those who do (obesity-prone).[6]

Overfeeding significantly attenuates fMRI activation in the visual cortex of obesity-resistant individuals compared with obesity-prone ones, meaning that signals that operate to reduce food rewards and thus reduce intake in lean individuals are not present in those prone to gain weight. The practical applications of these observations are not certain, but manipulation of their function by medications or even brain stimulation/ablation could be possible in the future as a means of curing obesity.

Can obesity be inherited? Those who argue against this notion claim that altering our genes takes many generations and that the obesity epidemic is a fairly recent phenomenon; therefore, not enough time has gone by to change our genes. Those in favor state that the development of "energy-thrifty genes" occurred when humans had less food, and now that we have an abundance of it, these genes favor the storage of fat and make us gain weight. Studies show that genetic factors may be responsible for 50%–80% of weight variations.[7] About 5% of cases of obesity are monogenic— that is, caused by single gene defects (11 different genes identified so far). In polygenic obesity, more than 100 candidate genes have been identified. Genome-wide association studies have consistently pointed to the *FTO* gene as being a main culprit in obesity, and because it is a highly conserved gene, it is passed to subsequent generations. This gene leads to production of a protein that is predominantly expressed in the hypothalamus, so its absence may lead to the non-homeostatic circuit overriding the homeostatic one. *FTO* gene polymorphism results in increased food intake in children and loss of control over eating, and *FTO* polymorphism carriers do not respond well to diets. What is more, *FTO* is also associated with diabetes independent of weight.

The second most common gene to be associated with obesity is called *transmembrane protein 18*. The polymorphism of this gene leads to weight gain and increased waist circumference and, even when adjusted for weight, also carries a higher risk of diabetes.

Deletion of the *SH2B adaptor protein* (located in 16p11.2) leads to resistance to leptin and obesity and, once more, insulin resistance.

Again, this gene is expressed in the hypothalamus and is capable of overriding homeostatic mechanisms. *Neuronal growth regulator 1* (1p31.1) is also highly expressed in the brain, and when absent, weight gain, larger circumference, and diabetes ensue. Both of these latter genes also modulate the growth of adipose cells. The hypothalamus produces something called an agouti-related peptide, which is an antagonist of the *melanocortin-4 receptor,* which, if activated, decreases food intake. Homeostatic mechanisms in the hypothalamus block these receptors, increasing food intake, and if the gene that encodes for them is deficient and the receptors absent, satiety does not occur, leading to extreme eating, decline in energy use, and obesity.

Obese children may also have a genetic defect that makes them eat more carbohydrates. Animal studies show that mothers fed a long-standing high-fat diet produce offspring who demonstrate increased adiposity, glucose intolerance, and altered brain appetite regulators.[8] Even in the face of only mild maternal over nutrition, these traits persist. It is hoped that knowledge of these gene defects will lead to personalized weight management and prediction of obesity and even perhaps gene manipulation in individuals at risk and that the effects of drugs may be monitored with fMRI.

However, as always, things are not that easy, and epigenetic factors may further alter the functions of these genes. Ingestion of monounsaturated fats changes the way many of these genes act.

So, do these genes change us or do we change these genes? Maybe both.

Once obesity is established, even in absence of diabetes, it increases arterial disease such as atherosclerosis. In one study, children with risk factors that included obesity had increased atherosclerosis progression in adulthood.[9] Because 4%–6% of all US children are obese, the neurologic implications of these findings are important. Unfortunately, there is no easy fix because lifestyle modifications, behavior treatments, and even medications are only minimally effective and most participants remain obese after completion of these treatments.[10] In obese children, the carotid arteries become thicker and stiff, and plasma markers of endothelial activation and injury are high. As the arteries stiffen, they cannot dilate to accommodate increased flow and the brain may not get enough blood when higher demand is in order.

Low back pain, the most common indication for lumbar spine MR imaging studies, is also correlated with weight. In one study in which participants were followed for 11 years, low back pain was either present at the beginning or developed during the study independent of other factors such as education, physical activity, and smoking.[11] Neurosurgeons know that physical therapy and surgery commonly fail when the lumbar spine of obese patients is operated on. Infections and re-operation rates are also higher in the obese.

A newly recognized and significant risk factor for back pain in the obese is metabolic syndrome. Metabolic syndrome is associated with a special type of weight: a large waistline (individuals with excess fat in their abdomen but relatively little elsewhere).

Other conditions associated with it are diabetes, high blood pressure, and high lipids. Overall, this syndrome is present in up to 20% of the adult US population and is highest in Hispanics. The prevalence of the syndrome is about 5% in those with a normal weight, nearly 60% in the obese, and nearly 39% in those with low back pain.[12],[13] Tumor necrosis factor (*TNF*) is produced by individuals with the metabolic syndrome and is known to cause low back pain; when it is blocked, the pain disappears. Furthermore, aortic atherosclerosis associated with metabolic syndrome has been linked to degenerative disk disease and low back pain.

The syndrome is also known to cause silent cerebral infarctions.

It is interesting that some investigators postulate that metabolic syndrome originates in the brain due to alterations in our circadian clocks. Normally, during sleep, our brain prepares our body for the next day's physical activity, but in modern life in which physical activity is minimal, this mechanism has been disrupted.

The hypothalamus releases hormones and alters the function of the autonomic nervous system, resulting in changes in blood pressure, insulin, abdominal fat breakdown, and glucose uptake, but all of these activities are no longer needed as we sit at our desks all day long. The more abdominal fat we have, the greater the amount of adipokines we produce. Adipokines are cell-signaling proteins secreted by fatty tissues that have immunomodulating capacities, TNF-_ being one of the most important ones. Additionally, adipose tissues produce hormones (called adipose- derived hormones), and their production becomes abnormal in patients with metabolic syndrome. One of these hormones is leptin, which as we saw above, can affect food intake. Obese individuals produce too much leptin, but instead of decreasing hunger, their brains become resistant to leptin and they just eat more.

There is a popular belief that up until the 1900s, fat was seen as attractive, that in women it signified health and the ability to have babies, while in men, it meant prosperity. Newer research shows that most pre-Victorians and others before were thin, and their diets, nutritious.[14] They ate many fruits and vegetables (mostly organic) and fiber, a diet akin to what we now call Mediterranean eating. No, they were not malnourished characters from a Dickens novel. They were actually healthy, and their physical activity is said to have been 3–4 times as much as ours.[15] Recent evidence suggests that back then, life expectancy was not much different from now, the incidence of degenerative disease was 10% of ours, and cancer was basically nonexistent (of course, infections were rampant and childbirth fatalities and accidents were common).

By the mid-Victorian times, diet and health had deteriorated significantly (cheap sugar, salted meats, and vegetable oils are just 3 popular products from the Agricultural Revolution responsible for obesity). The year 1900 was probably the last time we were a lean human race. Coming back to where I started, obesity was basically unknown in pre-Columbian Mesoamerica where the diet was gluten-free, low-carb, nutrient attenuated, and high in protein and fiber. Unfortunately, it is now in Mesoamerica where obesity is more prominent.

* The parabrachial nucleus complex (unknown to many neuroradiologists) is located at the junction of the midbrain and pons at the level of the superior cerebellar peduncle and is involved in the transmission of gustatory impulses.

NB: For those who are interested in this topic, this is a very nice article: Caballero B. **The global epidemic of obesity: an overview.** *Epidemiol Rev* 2007;29:1–5

REFERENCES

1. Food and Agriculture Organization of the United Nations. The nutrition transition and obesity. http://www.fao.org/focus/e/obesity/ obes2.htm. Accessed January 22, 2014
2. LATINOVOICES.Mexico obesity rate surpasses the United States', making it fattest country in the Americas. http://www.huffingtonpost.com/2013/07/09/mexico-obesity_n_3567772.html. Accessed January 22, 2014
3. Flegal KM, Carroll MD, Kit BK, et al. **Prevalence of obesity and trends in the distribution of body mass index among US adults, 1999–2010.** *JAMA* 2012;307:491–97
4. Centers for Disease Control and Prevention. Adult obesity facts. http://www.cdc.gov/obesity/data/adult.html. Accessed January 22, 2014
5. Passel JS, Cohn D. **US population projections: 2005–50.** *Pew Research Hispanic Trends Project*. February 11, 2008. http://www. pewhispanic.org/2008/02/11/us-population-projections-2005-2050. Accessed January 28, 2014
6. Grill HJ, Skibicka KP, Hayes MR. **Imaging obesity: fMRI, food reward, and feeding.** *Cell Metabolism* 2007;6:423–25
7. Moleres A, Martinez JA, Marti A. **Genetics of obesity.** *Curr Obes Rep* 2013;2:23–31
8. Rajia S, Chen H, Morris MJ. **Maternal overnutrition impacts offspring adiposity and brain appetite markers: modulation by postweaning diet.** *J Neuroendocrinol* 2010;22:905–14
9. Juonala M, Viikari JSA, Kanonen M, et al. **Life-time risk factors and progression of carotid atherosclerosis in young adults: the cardiovascular risk in young Finns study.** *Eur Heart J* 2010;31:1745–51
10. Kelly A, Barlow SE, Rao G, et al. **Severe obesity in children and adolescents: identification of health risks, and treatment approaches— a scientific statement from the American Heart Association.** *Circulation* 2013;128:1698–712

11. Heuch I, Heuch I, Hagen K, et al. **Body mass index as a risk factor for developing chronic low back pain: a follow-up in the Nord-Trondelag health study.** *Spine* 2013;38:133–39
12. Park YW, Palaniappan L Heshka S, et al. **The metabolic syndrome: prevalence and associated risk factor findings in the US population from the third national health and nutrition examination survey, 1988–1994.** *Arch Intern Med* 2003;163:427–36
13. Ha JY. **Evaluation of metabolic syndrome in patients with chronic low back pain: using the fourth Korea national health and nutrition examination survey data.** *Chonnam Med J* 2011;47:160–64
14. Westcott P. Healthy eating Victorian style. Saga. http://www. saga.co.uk/health/healthy-Eating/healthy-eating-victorian-style. aspx. Accessed January 30, 2014
15. Clayton P, Rowbotham J. **How the mid-Victorians worked, ate and died.** *Int J Environ Res Public Health* 2009;6:1235–53

THE BENEFITS OF BEAUTY

Beauty is only skin deep, but the ugly goes clean to the bone.

Dorothy Parker

Although many say that beauty is in the eye of the beholder, I do not agree. To me, beauty is clearly defined, objective, and even measurable. Someone considered beautiful will be seen that way by most observers, even those from different cultural and ethnic groups. The opposite is also true: An ugly person is considered as such by nearly everyone. When one looks at a female face, the following features are associated with beauty: narrower face and nose, less fat, full lips, minimal hypertelorism, longer eyelashes, and high cheek bones; for men, the same plus darker eyebrows, broader upper half of the face, prominent mandible and chin, and no wrinkles between the nose and corners of the mouth.[1] Of course, to these mixtures you can add tan skin, which is considered more attractive nowadays than paleness.* Facial symmetry is also a hallmark of beauty but possesses a tricky conundrum.

Symmetric faces are generally more beautiful, but perfect symmetry may not be beautiful. If one creates chimeric faces by duplicating either the right or left sides, some faces become beautiful while others turn ugly. Curiously, individuals with symmetric faces have overall better health than those with asymmetric faces.[2] Could it be that better genes are reflected by beauty and health? It seems so, because mate quality is generally based on those features—that is, to select a healthy/beautiful partner you, too, must have these characteristics that, in turn, will lead to your children having 2 sets of healthy and gorgeous genes. Not only do individuals choose mates who resemble themselves, after 25 years of marriage they become even more alike (advice: try to find a beautiful companion).[3] There is now a "dating" Web site that allows you to find someone who shares your facial features, a fact touted to increase compatibility.[4]

Mother Nature and Father Time are beauty's greatest enemies; the age at which most men find women's faces most attractive is 24.8 years.[5] Beauty ratings in relation to age affect women more than men, and beauty ratings for both genders drop with advancing age. Curiously, ugly persons are less affected by age, meaning that once ugly, they remain ugly for life, and that seems to overcome the effects of aging. Urban British men now spend more time and effort in their daily grooming than women (83 versus 79 minutes).[6] In the United States, where men are still men, the overall time and ratio are different (32 versus 44 minutes). With age, both genders spend more time in this activity, and the face gets the most attention. After all, as the popular and famous neurologist Oliver Sacks put it, "It is the face, first and last, that is judged 'beautiful' in an esthetic sense."[7]

The reality is that human beauty is scarce, and like anything else scarce, it becomes a commodity. Thus, the success of plastic surgery should not surprise anyone. Eyelid surgery is the third most commonly performed cosmetic surgery (after breast augmentation and liposuction), and the top 5 noninvasive procedures are done for the face (botulinum and hyaluronic acid injections, laser hair removal, skin resurfacing, and chemical peels).[8] I find it surprising that the highest rate of cosmetic surgery occurs in Korea, followed by Brazil, Taiwan, and the United States.[9] Plastic surgery done to alleviate the anxiety of aging simply does not make one look younger; you look...well, just like you had surgery.

Cosmetic surgery done to improve an asymmetry such as a crooked nose does, however, make one better-looking. Spending more on clothing, haircuts, and makeup does little to increase beauty. For every US dollar spent on these items, one can expect a return of only 4 cents.[10] From a purely financial point of view, spending on education is a better deal as each extra year of schooling translates into a 10% increase in wages. Remember that on average, women value education in men more than looks. The reverse is not true. Would this change if the earning capacity of women was on par with that of men?

In his entertaining book, *Beauty Pays: Why Attractive People are More Successful,*[11] Daniel Hamermesh addresses the economics of beauty. From it, the reader can clearly see that physical beauty _ money. This is true not only in professions such as high-fashion modeling and sex work where gains are directly related to beauty, but applies to nearly all jobs. Regardless of physical beauty, it seems that immigrants in any country earn less than others as their looks are generally considered different from those of natives. The difference in wages between beautiful and ugly individuals is not much, about 4%–5% but becomes significant over a lifetime. Social networking, such as Facebook has made beauty even more important. Appreciation of beauty does not vary with intelligence, but too much beauty is sometimes penalized (the so-called "bimbo factor"). The concept of beauty extends to the body, and short and obese individuals have smaller salaries than their taller and slimmer counterparts.

Neuroesthetics, a subspecialty that first appeared in 2002, attempts to study the neural bases for the contemplation and creation of beauty. Emotion plays an important role in the appreciation of beauty, and what else makes us more emotional than a beautiful person? Makeup is nothing more than a means of exaggerating features already present and results in higher brain activity in certain locations. One of the 8 laws of artistic experience of Ramachandran and Hirstein is symmetry.[12] To perceive a human being as artistically pleasurable, symmetry must be present. Evolutionary biologists state that asymmetry is associated with infection and disease.[12] Symmetry results in activations in the parietal and premotor regions subserving spatial processing.[13] fMRI signals are seen in the frontomedial cortex and intraparietal sulcus when one is exposed to beauty. The perception of symmetry and beauty relies on brain areas that support high-level visual analysis.

Moreover, areas associated with judgments are also activated, and appreciation of beauty cannot occur without these. In 1 experiment, the same brain region (the medial orbitofrontal cortex) was activated by both musical and visual beauty.[14] As the famous Irish statesman and philosopher Edmund Burke said, "Beauty is, for the greater part, some quality in bodies acting mechanically upon the human mind by the intervention of the senses" (*On the Sublime and Beautiful*).

It is obvious that nature and economics favor beautiful individuals, so what about the rest of us? California, the District of Columbia, and 5 cities prohibit discriminatory hiring based on looks (pictures are never requested when applying for jobs in those regions but can be voluntarily submitted). Laws in Michigan and San Francisco explicitly forbid taking into account height and weight when hiring. A recent article in *The New York Times* goes as far as proposing that ugliness should be protected by extending the Americans with Disabilities Act.[15] Because beauty is inherited, it could be argued that ugliness is already protected by the Genetic Information Nondiscrimination Act passed in 2008, which outlaws discrimination in employment and health insurance based on genetic data.[16] The overall impact on the economy would be small as only 1%–2% of the population is truly ugly, but who would be willing to be officially labeled as "ugly"? If that denomination compensates for the US $230,000 per individual lost in a lifetime, I believe that many would.

In academics, we prize intellect over beauty, but the reality is otherwise. Beautiful professors attract more students at colleges, while ugly ones face empty or nearly empty classrooms, a worldwide valid observation. The Web site http://www.ratemy professors.com allows users to input opinions not only on academic endeavors but also about how their teachers look. "Hot" professors make, on average, 6% higher wages than their not-so hot counterparts. This observation applies to many other professions, including football players and CEOs of the largest Fortune 500 companies. Charitable organizations know this well; their most beautiful solicitors generally bring in more donations.

Handsome public attorneys are more likely to move into the private sector later in life than ugly ones. Attractive political candidates get twice as many votes as their opposites. In economics, this is known as "customer discrimination"—that is, a good-looking person becomes a part of what the customer is buying. The store chain Abercrombie & Fitch is well-known to hire, promote, and use in publicity only individuals they rate as "hot," a behavior that brought a large lawsuit based on its discriminatory policies.[17] Of course, that has not changed the behavior of many similar businesses— just walk into one of these stores and you will find yourself surrounded by beautiful, mostly female, employees in their late teens or early 20s.

Customers may, more concretely, buy beauty. An Internet search that I did for purposes of this Perspectives revealed several advertisements by "beautiful and clever" women selling their eggs to infertile couples. In Great Britain, where a standard egg donation goes for about £250, beautiful women's eggs may fetch up to £12,000.[18] The Web site http://www.beautifulpeople.com went as far as offering sales of sperm from handsome men, and in 2010, its CEO expelled 5000 members for being ugly.[19] Because mixing 2 sets of genes generates an enormous amount of randomness and variability, good looks cannot be completely guaranteed if one buys these eggs and sperm. Evolution selects the best-looking individuals, and why not? On average, they have better lifestyles, earn more, and are healthier, physically and mentally. There are still great jobs where physical beauty is not critical, but they are becoming scarce. Being the editor of a scientific journal is one of them, but I wonder, if I were better-looking would *AJNR* have more subscribers?

* Factoid: Before the 1920s, tan skin was considered "lower class" and borne mainly by agricultural workers. Soon after Coco Chanel started vacationing in the French Riviera, tanning became associated with the leisurely lifestyle of the rich and famous and seen as beautiful.

REFERENCES

1. Characteristics of beautiful faces. http://www.uni-regensburg.de/Fakultaeten/phil_Fak_II/Psychologie/Psy_II/beautycheck/english/ prototypen/prototypen.htm. Accessed February 1, 2012
2. Rhodes G, Zebrowitz LA, Clark A, et al. **Do facial averageness and symmetry signal health?** *Evol Hum Behav* 2001;22:31–46
3. Zajonc RB, Adelmann PK, Murphy ST, et al. **Convergence in the physical appearance of spouses.** *Motiv Emot* 1987;11:335–46
4. Find Your FaceMate. http://www.findyourfacemate.com. Accessed February 1, 2012
5. Genetics and science behind beautiful people and physical attraction. http://www.healthfiend.com/beauty/genetics-and-science-ofbeauty-attractive-handsome. Accessed February 1, 2012
6. Mail OnLine. Rise of the metrosexual: men now spend longer getting ready to go out than women. http://www.dailymail.co.uk/femail/article-1249709/Rise-metrosexual-Men-spend-longer-gettingready-women.html. Accessed February 1, 2012
7. Sacks O. **Face blind.** *The New Yorker.* August 30, 2010:36–43
8. Top five procedures. http://www.surgery.org/sites/default/files/2010-top5.pdf. Accessed February 1, 2012
9. asianplasticsurgery. Asian plastic surgery guide. http://www.asian plastic-surgeryguide.com. Accessed February 1, 2012
10. Lee B. Plastic surgery and attitudes of beauty and success. http://www.chinadaily.com.cn/english/doc/2004 – 07/05/content_345598.htm. Accessed February 1, 2012
11. Hamermesh D. *Beauty Pays: Why Attractive People are More Successful.* Princeton, NJ: Princeton University Press; 2011
12. Ramachandran VS, Hirstein W. **The science of art: a neurological theory of the aesthetic experience.** *J of Consciousness Studies* 1999;6:15–51
13. Jacobsen T, Schubotz RI, Hofel L, et al. **Brain correlates of aesthetic judgment of beauty.** *Neuroimaging* 2006;29:276–85

14. Ishizu T, Zeki S. **Toward a brain-based theory of beauty.** *PLoS ONE* 2011;6:1–10
15. Hamermesh D. **Ugly? You may have a case.** *The New York Times Sunday Review*. August 27, 2011. http://www.nytimes.com/2011/08/28/opinion/sunday/ugly-you-may-have-a-case.html. Accessed February 1, 2012
16. Public Law 110 - 233 - Genetic Information Nondiscrimination Act of 2008. http://www.gpo.gov/fdsys/pkg/PLAW-110publ233/contentdetail.html
17. AFjustice.com. $50 million, less attorney's fees and costs, paid to class members in December 2005 in Abercrombie & Fitch discrimination lawsuit settlement. http://www.afjustice.com. Accessed February 1, 2012
18. Connell C. I'm beautiful, clever, and I'll sell you my eggs for £12,000. http://www.dailymail.co.uk/femail/article-1206660/Im-beautifulclever-Ill-sell-eggs-12– 000.html. Accessed February 1, 2012
19. Eherlich B. Dating site launches online sperm and egg bank for "beautiful people." http://mashable.com/2010/06/__

DRESS FOR SUCCESS

> *It is impossible to wear clothes without transmitting social signals.*
> *Every costume tells a story, often a very subtle one, about its wearer.*
>
> Desmond Morris in *A Field Guide to Human Behavior*

A couple of years ago, I visited the Mayo Clinic in Rochester, Minnesota, and was surprised to see that most physicians still wore suits, while here in Chapel Hill where I work, I am only one of a handful to do so.* I wear a suit because I feel that it makes for easy choices in the morning, looks respectful and elegant, and after a certain age (mine) it flatters the aging body. The "business suit" (sometimes also called a "lounge suit") appeared in the late 19th century, and typically all of its pieces (2 or 3) are made of the same cloth. Suits tend to compensate for physical variations by having no indication of a waistline and by adding padding to the shoulders; minor variations such as lapel width go in and out of style regularly. Business suits are a type of uniform, and today uniforms are worn mainly by 3 professions: the armed forces, commercial aviation, and medical staff.

Smaller groups of individuals such as sports professionals and hygiene workers and even some students in private schools still wear uniforms. Commonly seen in medicine are 2 types of uniforms: white coats (or laboratory coats) and scrubs.

Most lab coats are knee length and have long sleeves to offer the most protection. Additionally, they are usually made of cotton because of its high capacity to absorb liquids. Professionals for whom sleeves prove uncomfortable or prone to contamination, such as microbiologists and pharmacists, prefer short-sleeve smocks. In a survey of nearly 300 doctors, only 1 in 8 wore a white coat, despite the fact that over 50% thought they should.[1] Specialists who are the least likely to wear white coats include psychiatrists and pediatricians, while those more likely to do so are surgeons and gynecologists.

Older physicians are also more likely to wear them than younger ones.

Scrubs were designed for surgical personnel but are now worn by nearly anyone who works in health care (they are also mandated in some prisons). Similar to lab coats, scrubs were initially white, but with the advent of modern operating room illumination, that color resulted in eye strain. During the 1950s and 1960s, most hospitals adopted green scrubs in an attempt to lessen this strain.

Today scrubs vary in color, often distinguishing among different specialties, and those used by pediatricians may have cartoon characters printed on them.[2] Because they are very comfortable, they are commonly used outside the hospital as pajamas, for working out, or just hanging out at home (it does upset me when I go to bar or restaurant and see hospital personnel in scrubs there). I also find it unfair to take the hospital germs home and then bring those from home to the hospital.

Wearing a uniform, such as a white coat, alters the perception of those who see us and also affects the way the wearer thinks. In an experiment, students wearing white coats noticed twice as many errors during a test compared with those wearing street clothes. They also did better at spotting differences during tests designed to measure sustained attention.[3] British patients prefer their male physicians in a suit and tie but their female physicians in white coats.[4] In Great Britain, the favorite male attire seems to be a tweed sports jacket and a tie and informal shirt. Overall, 64% percent of patients think that the way their physicians dress is very important, and 41% say they their confidence in their physicians' abilities is based on appearances. Casual dressing for doctors is a bad tactic: It decreases perceptions of authority regardless of sex, paradoxically decreases perceptions of friendliness and trust, and also lowers attractiveness.[5] There are sex differences in perception, and female patients prefer their physicians in white coats, while male patients prefer them formally dressed. Thus, the best attire may be a white coat over formal wear and removal of the white coat when the occasion calls for it. The attitude toward physicians' dressing styles may be age-related. In a different study, 43% of teenagers responded that the way their physicians dressed made no difference and only 26% preferred them in white coats.[6]

Patients prefer their anesthesiologists to wear the traditional business suit and find blue jeans an undesirable choice.[7]

A different study, also involving anesthetists, found that patients had a preference for name tags, short hair, and white coats but disapproved of clogs, jeans, sneakers, and earrings.[8] In a study done in Italy, patients also preferred formal dressing and name tags.[9] This very complete study of patient preferences also rated the following as very highly favorable: short nails, well-kept teeth, and light makeup on females. The following received very low ratings: tattoos, piercings, obesity, sandals, long hair on men, and heavy makeup on women. In another large study done in South Carolina, all respondents preferred their physicians in white coats, followed by business and surgical (scrubs) attire, and last casually dressed (jeans and polo shirts).[10] In the military where uniforms are standard, patients prefer their doctors in white coats over scrubs.[11] In that same environment, nonwhites and Hispanics had higher rates of preference for more formal dress than other groups.

Not only do most patients show similar preferences but their parents do, too. A study performed at the Children's Hospital in Cincinnati showed that parents prefer and express more confidence in physicians who are formally dressed than in those without a white coat, no necktie, and tennis shoes. These preferences were independent of the severity of the illness, time of visit, insurance group, race, and sex.[12] Some studies point out that white coats may not be completely innocuous to patients. The "white coat syndrome" is a phenomenon in which patients exhibit high blood pressure when facing someone dressed in a white lab coat.

Female students at an American university were asked to rank the same teachers dressed in traditional business and Indian attire, and not surprisingly, the former received a more positive evaluation.

[13] Although females prefer their teachers formally dressed, mixed-sex college classes express positive ratings for instructors dressed in casual clothes (jeans, t-shirts, flannel shirts).[14]

When I was a resident in radiology, the dress code was always a lab coat, a necktie for men, and no scrubs or tennis shoes except when doing procedures. If we arrived at the hospital without a tie, the program director would take us to his office where he had a drawer full of ties from which we chose what to wear that day (I recall that most were ugly, perhaps an incentive not to forget them again). Neckties of different types have been worn since Roman times, but it was not until 1926 that the current necktie (the long one) was designed and became popular. Studies show that about 20% of neckties are contaminated by the third hour worn, often with antibiotic-resistant micro-organisms (the rate of contamination for lab coats is about 25%, more if they contain polyester).

[15] Bow ties are also prone to contamination though less often than long ties and because both are seldom washed, colonization with micro-organisms may remain in them forever. Because of this, some companies have started to apply nanoparticles that presumably "lock" the silk fibers found in neckties, preventing bacteria from getting into them. In one study, these so-called "safety ties" were found to have more bacteria than regular ties!

Counterintuitively, it was the knots of the ties and not their tips that contained more bacteria (perhaps because we tend to adjust them often).[16] If one is going to wear a tie, it is better to wear a bow tie and change it every day.[17]

White coats also allow one to carry stuff in the pockets. In one study, 70 individuals were asked what their lab coat pockets contained.

These were all clinicians, and regardless of their status, stethoscopes, pocket manuals, "to do" lists, and telephone numbers were the items most commonly found in their pockets. Faculty and older staff were not fond of carrying handouts from lectures that were popular among younger staff. Students and faculty both carried family pictures. When asked about the usefulness of these items, obviously medical equipment received higher scores, while lecture handouts were considered less useful than family photographs.[18]

Wearing perfume or cologne at the hospital or clinic, whether you are the physician or the patient, is generally discouraged. A number of workplaces are adopting fragrance-free or scent-reducing policies. It is thought that strong smells may induce allergies or asthma and are frowned on by most hospital safety committees (taking this to an extreme, Harrison Medical Center in Bremerton, Washington, accepts only flowers that are "less" fragrant for their patients). In reality, most allergies are induced by specific proteins that fragrances do not contain because what gives them their odor is actually a series of volatile hydrocarbons, which are not known to stimulate the immune system but, in large concentrations, may result in chemical irritations.[19] Anyway, because most soaps, shower gels, deodorants, and body lotions contain some fragrance nowadays, wearing perfume is less popular (reflected by the decreasing revenues of perfume/cologne sales in the last few years and the fact that most of today's perfumes and colognes are very mild).

For the time being, I will continue wearing a business suit on most days at work. I generally do not wear a white coat because I do not see patients, but as the evidence indicates, if you are consulting with patients it is a good idea to wear one and to change it often to keep it clean.

* The Mayo Clinic encourages their physicians to wear business suits rather than white coats because suits are thought to convey professionalism and expertise.[20]

REFERENCES

1. BBC News. Doctors should wear white coats. http://news.bbc.co.uk/ 2/hi/health/3706783.stm. Accessed March 4, 2014
2. Wikipedia. Scrubs (clothing). http://en.wikipedia.org/wiki/Scrubs_%28clothing%29. Accessed March 4, 2014
3. Adam H, Galinsky AD. **Enclothed cognition.** *J Experimental Social Psychology* 2012;48:918–25
4. McKinstry B, Wang JX. **Putting on the style: what patients think of the way their doctor dresses.** *Br J Gen Pract* 1991;41:275–78
5. Brase GL, Richmond J. **The white-coat effect: physician attire and perceived authority, friendliness, and attractiveness.** *J Appl Soc Psychol* 2004;34:2469–81
6. Neinstein LS, Stewart D, Gordon N. **Effect of physician dress style on patient-physician relationship.** *J Adolesc Health Care* 1985;6:456–59
7. Sanders LD, Gildersleve CD, Rees LT, et al. **The impact of the appearance of the anesthetist on the patient's perception of the pre-operative visit.** *Anaesthesia* 1991;46:1056–58
8. Hennessy N, Harrison DA, Aitkenhead AR. **The effect of the anaesthetist's attire on patient attitudes: the influence of dress on patient perception of the anaesthetist's prestige.** *Anaesthesia* 1993;48:219–22
9. Sotgiu G, Nieddu P, Mameli L, et al. **Evidence for preferences of Italian patients for physician attire.** *Patient Prefer Adherence* 2012;6:36–67
10. Rehman SU, Nietert PJ, Cope DW, et al. **What to wear today? Effect of doctor's attire on the trust and confidence of patients.***AmJ Medicine* 2005;118:1279–86
11. Lund JD, Rohrer J, Goldfarb S. **Patient attitudes toward the use of surgical scrubs in a military hospital clinic.** *Patient Prefer Adherence* 2008;2:85–88
12. Gonzalez Del Rey JA, Paul RI. **Preferences of parents for pediatric emergency physician's attire.** *Pediatr Emerg Care* 1995;11:361–64

13. Chowdhary U. **Instructor's attire as a biasing factor in students' ratings of an instructor.** *Clothing & Textiles Research Journal* 1988;6:17–22
14. Morris TL, Gorham J, Cohen SH, et al. **Fashion in the classroom: effects of attire on student perceptions of instructors in college classes.** *Communication Education* 1996;45:135–48
15. Abuannadi M. **Should physicians be banned from wearing neckties in medical venues?** *General Surgery News* 2011;38:4. http:// www.generalsurgerynews.com/ViewArticle.aspx?d_In%2Bthe%2 BNews&d_id_69&i_April%2B2011&i_id_719&a_id_16975. Accessed March 4, 2014
16. Bosch W, Hedges MS, Cawley JJ, et al. **Do nano-treated neckties reduce the carriage of bacterial pathogens from neckties of physicians?** In: *Proceedings of the 48th Joint Annual Interscience Conference on Antimicrobial Agents and Chemotherapy and the 46th Infectious Diseases Society of America Annual Meeting*, Washington, DC. October 25–28, 2008
17. Biljan MM, Hart CA, Sunderlad D, et al. **Multicentre randomized double bind crossover trial on contamination of conventional ties and bow ties in routine obstetric and gynaecological practice.** *BMJ* 1993;307:1582–84
18. Lynn LA, Bellini LM. **Portable knowledge: a look inside white coat pockets**. *Ann Intern Med* 1999;130:247–50
19. Senger E. **Scent-free policies generally unjustified.** *CMAJ* 2011;183: E315–16
20. Berry L, Bendapudi N. Working Knowledge for Business Leaders. Clueing in customers: why docs don't wear white coats or polo shirts at the Mayo Clinic. http://humanresources.ku.edu/sites/sld.ku.edu/files/docs/2014_summit/Clueing_in_customers.pdf

DO YOU NEED A COACH? I DO

I recently reached an anniversary of sorts; during the 2015 Symposium Neuroradiologicum in Istanbul, Turkey, I gave my

900th invited lecture. Those who were there witnessed what I think were 2 unexceptional lectures. Why? I practiced each about 10 times before the meeting and thought I knew them well. The truth is that sadly, I am not a natural speaker. When I think about gifted speakers, Drs. Thomas Naidich and Anne Osborn immediately come to mind. We participate in several events every year, and many times I have seen Dr. Naidich sitting in the first row, building his next conference and immediately thereafter delivering it flawlessly. I will never be able to do that; for me "practice makes perfect" and sometimes as in the Symposium, it does not.

Because I think that I know my conference topics well, perhaps my delivery needs work. Maybe I talk too fast (yes, I am certain of that), maybe my body posture needs adjustment (not sure about this but gesticulating does keep nervousness at bay), or maybe I do not make enough eye contact with the audience. A million things can go wrong, but how to improve the most urgent ones is not clear to me. Would coaching help? I think that it would be helpful if someone were to film me and then constructively criticize how my lectures went. When one lectures as much as I do and feels the responsibility of representing one's institution, journal, and professional societies, one worries about the usual stuff public figures do, but the difference is that most (and other not-so-public figures) have coaches who help them refine their deliveries and image.

In his essay "Personal Best," Dr Atul Gawande begins with the following quotation (credited to Barry Blitt, an author and illustrator for the *New Yorker* magazine): "No matter how well trained people are, few can sustain their best performance on their own.

That is where coaching comes in."[1] Dr Gawande stated that after 8 years as a surgeon, his performance in the operating room (OR) reached a plateau, so he decided to try a coach. He contacted a surgeon he admired and asked him to evaluate his OR behavior. The coach pointed out several needed improvements, and once implemented, Gawande asked to be re-assessed.

He also recorded his operations and later watched them with his coach. Of course, I imagine Gawande to be very self-assured and not easily hurt by criticisms. However, regardless of how he felt at the time of his coaching, he concluded that his OR skills improved.

If surgeons, athletes, musicians, singers, chess players, and public speakers, among others, have coaches, why not we radiologists?

Today, excellence reigns among musicians, and the word "genius" has lost most of its importance. Most musicians and most professional athletes are excellent at what they do, and the difference between the excellent and truly great is not visible to the untrained like me. Coaching is about self-improvement and achieving perfection; thus, it may be a lifetime activity for the coach and coachee. Coaching strives to make us better without the addition of drugs, implants, and other "enhancements." Coaching is also a highly specialized activity often requiring not 1 individual but a team of professionals. For athletes, their coaching teams are formed by scientists, physicians, nutritionists, administrators, journalists, engineers, stylists, and many more.[2]

Because winning in sports is nowadays a matter of milliseconds, every little bit counts. Coaching also prevents bad habits from forming and those already there from becoming routine, but coaching is not mentoring. Mentoring is defined as "the relationship between an older more experienced individual with a younger less experienced person with the goal of developing the career of the latter."[3] Mentoring seldom involves payments, while coaching does, and I think that mentoring is a much more complex and difficult relationship (not to say overall less successful).

Let me now discuss some elements of coaching. The International Coaching Federation (http://www.coachfederation.org/) counts over 20,000 members (coaches) and offers several levels of certification (associate, professional, master, and so forth). Most of its activities, but not all, are at the executive levels and have spun considerable data. What follows is a summary of some of that literature (in the financial world, performance and profitability before and after coaching are easily measured indicators).

Coaching is akin to psychotherapy as far as human contact goes, but it differs from it by being highly focused and concentrating on the present and future rather than the past.[3] While a patient generally pays a therapist, in coaching, an organization generally handles the costs, thus inserting at least a third person into the relationship.

Both coaching and psychotherapy, however, try to change behavior. Coaching results in improved performance, commitment, efficacy, and leadership. An effective coach-coachee relationship depends on good mutual communication, trust, collaboration, and commitment to the process. After an initial observation stage, the coach provides feedback on performance and potential. These evaluations continue throughout the coach-coachee relationship and thus constantly serve to refine it.

The second stage is implementation, and though the coachee ceases to be under constant observation, he or she continues to meet regularly with the coach to discuss obstacles and successes, thus learning to exert some control over the items that need improvement. As if all of this is not complex enough, we need to bring back into the scene the third person previously mentioned. A coach-coachee relationship in which a supervisor (or, in general, the working environment) is not actively invested in the process is bound to fail. For coaching to succeed, the supervisor must believe from the start that it will and encourage both coach and coachee when improvements and changes are achieved.

Many academic radiology programs have some type of mentorship program, and from what I know about them, most are failures. With the help of American Society of Neuroradiology (ASNR) and the American Roentgen Ray Society (ARRS), this year we started 2 mentoring pilot projects trying to match senior and junior individuals in both societies. Although the final data are not in, the ASNR part has been only moderately successful and its ARRS counterpart less so. While one does not have to admire a coach, I feel that one must admire a mentor for the relationship to work. Self-choosing a mentor is critical, and that is why our pilots with ASNR and ARRS have not worked well (for the ASNR one, I attempted to match mentors with mentees). Therefore, here is my first proposal: Ask our larger organizations to establish coaching programs akin to what the financial institutions do. One or more senior radiologists could observe our senior residents, fellows, and attendings (at all levels) at work and coach them to become better radiologists. Alternatively, this could be done by using video capture. Real work situations could be evaluated for knowledge, safety, quality, education, collegiality, and efficacy and take the place of the complex general and specialty certifications and all that is needed to maintain them as valid. Scheduled visits, evaluations, and feedback could take the place of re-certifications; my feelings are that the current requirements to obtain these certifications really do not make us better radiologists because we all tend to keep up with knowledge, while at the same time, we design very poor-quality improvement programs, which in the end are of no use to anyone.

Coaching may also be useful beyond our clinical practices.

After my significant involvement with 2 scientific societies (ASNR and ARRS), I have come to perceive succession for their key positions as a great challenge. If we think of our nonprofit organizations as families, perhaps it would be helpful to look at how coaching helps in the succession of family businesses. Difficulties found in our organizations are similar to those of family-run businesses: competition, ego, and jealousy, just to mention a few. However, we also share many positive features: values, commitment, legacy, and a desire to survive. Using data from 630 family-owned and -run companies, a study concluded that coaching had a greater influence on the performance of these businesses than mentoring.

[4] In these situations, short and focused coaching was cheaper than mentoring and showed immediate changes in professional performance and skill development. Individuals who were coached also performed successfully when they took over those businesses. My second proposal is that once our larger scientific societies have identified those individuals who eventually will become their leaders, why not hire coaches for them? The current process of mentoring those individuals and making them spend years (sometimes decades) as members of boards and councils does not always work; we are all aware of many chosen nonprofit leaders who were not prepared for the jobs.

Now back to where I started. I do not believe that public speaking coaching is for everyone. If you only give occasional lectures and most are case presentations to your residents, there is no need to go overboard and try to find a coach. However, if you are, like me, delivering some 50 invited lectures per year and still unsure if you are doing it well, some help may be useful. Public speaking coaches will rapidly tell you 2 things: Lecturing is a fact of life for most people (especially educators), and yes, most people are afraid of it (admittedly I still get nervous sometimes but have not been able to identify what triggers it). I Googled "coaching for public speakers" and got more than 15 million hits. Reading the material found in the first 3 pages just confused me; many feel that coaching is essential (the opinion of most who sell those services), and some feel that it is not (these generally recommend recording your voice and image and self-coaching). The same did not happen when I Googled "vocal coaching" with emphasis on opera singers (they have consistently and reliably used coaches for a very long time). Opera coaching strives to extend vocal range and add projection to the voice (I like the part about adding "projection" because I feel that the voices of most academic lecturers do not project well), smooth out vocal wobbles and cracks (I have been trying to do this with the help of a glass of water on the podium), prepare for concerts, sing without straining, and other more technical stuff that does not apply to us (adding squillo, removing vibrato, and so forth) as well as recovering your voice after trauma. Some coaches (especially those residing in other countries) offer coaching via Skype. Coaching opera is difficult because the science of voice and the art of performance are intermingled.

Yet that is exactly what I do when I get up to the podium to lecture, a combination of voice and person. Voice and performance are the 2 main factors by which any public speaker is judged. To those, one has to add knowledge when judging our academic lecturers.

To me, there is no simple way of self-evaluating these 3 aspects, so does it make sense that at least some of us would benefit from coaching?

REFERENCES

1. Gawande A. **Personal best: top athletes and singers have coaches—Should you?** *New Yorker*. October 3, 2011. http://www.newyorker. com/magazine/2011/10/03/personal-best. Accessed November 19, 2014

2. Surowiecki J. **Better all the time: how the "performance revolution" came to athletics—and beyond.** *New Yorker.* November 10, 2014. http://www.newyorker.com/magazine/2014/11/10/better-time. Accessed November 19, 2014
3. Baron L, Morin L. **The coach– coachee relationship in executive coaching: a field study.** *Hum Resource Dev Q* 2009;20: 85–106
4. Utrilla Nunez-Cacho P, Torraleja FA. **The importance of mentoring and coaching.** J Manag Org 2013; 19: 386-404

A CALL TO ACTION: MAINTAIN YOUR HAPPINESS, BE GENEROUS!

Generosity comes in many flavors: the giving of time, resources, goods, and, of course, money. No matter what is given, Americans are considered to be most generous. With respect to donating money, in 2006 Americans donated over $295 billion, which, corrected for inflation and population changes, made for a 190% increase compared with 50 years previously.

Americans donate more than the citizens of any other country, encouraged by a tax system of deductions.[1]

"Generosity" is defined as the habit of giving without expecting anything in return. The practical manifestations of generosity are donations. In basically all religions, generosity is rewarded, and in Buddhism, it is the opposite of greed. Americans who donate give one-third to religious institutions, the rest to secular causes, and it is known that "religious and conservative" individuals donate more than nonreligious liberal individuals.

The University of Notre Dame's initiative, the Science of Generosity, explores the relationship between philanthropy, volunteerism, and altruism. This initiative gives about $3 million annually for research, and in its first year, 2009, it received over 600 proposals, of which 9 were funded. A list of currently funded projects can be found at: http://generosityresearch.nd.edu/ current-research-projects. An interesting project funded by this institution looked into the genetic origins of altruism in young children.[2] The authors found that the presence of the arginine vasopressin (AVP) receptor 1A leads to a lower altruistic proclivity.

In short, AVP serves as a peripheral hormone that regulates water balance and has effects in the hippocampi and amygdalae. It is also thought that AVP plays some role in the brain's dopaminergic mechanisms because its sites of expression are somehow related to those of the dopamine reward-associated pathways.

Oxytocin is also involved in mediating generosity. In one experiment, individuals injected with 40 IU of this hormone were found to be 80% more generous than those who received a placebo.[3]

This last study comes from the Center of Neuroeconomics Studies headed by Dr Paul J. Zak, a mathematician and economist with postdoctoral training in neuroimaging at Harvard. Dr Zak believes that oxytocin is our "moral molecule," and he has written a popular book about it.[4] It seems that oxytocin is also associated with feelings of well-being, and that is why individuals who give feel pleasure at doing so. fMRI shows activation of the precuneus and lingual gyri when generosity is called for, and greater lingual gyral activation is associated with an increased propensity to give.[5] The same brain regions are used when taking an outside perspective of one's self, thinking about the death of a loved one, and recalling vivid memories of one's life. These thoughts bring us closer to mortality, and the feeling presumably triggers a desire to leave a legacy and give.

What is fascinating is that generosity is its own reward because it results in additional oxytocin release and increasing feelings of happiness (not a "vicious" but a "virtuous" cycle). Whether we are being generous for our own benefit or acting that way on behalf of others, the brain activates and produces identical feelings of reward.

Happiness leads to generosity, and this idea is explored in Richard Powers' wonderful book *Generosity: An Enhancement.*[6]

In it, a young woman with excess happiness that leads to marked generosity is thought to harbor the gene for happiness/generosity and is exploited and, not surprisingly, abused by the media. Of course, such a condition does not exist, and in the book, it serves as a device to tackle the idea that some excessively generous individuals are viewed by society as anomalies. What is true is that some bipolar individuals express extreme happiness during their hypomanic periods, which disastrously leads to periods of mania and/or depression.[7] The "syndrome of excess happiness" may be a serious psychiatric condition.

Happiness is now more popular than ever. The Greater Good Web site (www.greatergood.com) contains more than 400 articles dealing with happiness and, specifically, how to bring up happy children. Extreme and constant happiness leads to decreased creativity.

[8] Extreme happiness also leads to riskier behaviors, such as binge eating, sexual promiscuity, and drug abuse. Children who are considered very happy have higher mortality rates because they tend to engage in riskier behaviors. The problem is also that the term "happiness" is really an umbrella that encompasses different types of feelings and not just one. What is even worse, obsessively pursuing happiness makes you unhappy.

Does having money (lots of it) make us happier and more generous? The best rated jobs in the United States are dentists and physicians; and though they are not the best paid ones, no one in these professions lacks money.[9] Although among physicians, radiologists are number 3 on salary scales, they do not even rank in the top 10 when job happiness is evaluated.[10] If asked, only 50% of all physicians stated that they would study medicine again. Judging by donations, specifically those to the Foundation of the American Society of Neuroradiology (ASNR), I cannot say that neuroradiologists are very generous.* An informal poll, taken by myself, tells me that neuroradiologists are not happy with their salaries (think they should get more) and especially are not happy with the idea of lower ones in the next few years. How much money is needed to be happy?

Daniel Kahneman, a Princeton psychologist who won the Nobel Prize for economics in 2002, has explored the issue of money and happiness. He has concluded that happiness and a sense of well-being increase with salary but just up to US $75,000 per year.[11] Above that amount, there are no more increases in happiness (however people making at least that amount are twice as happy as those making, on average, US $20,000 per year). He suggests that higher income buys satisfaction but not happiness.

Moreover, individuals earning higher incomes tend to be tenser, lose their ability to savor small pleasures, and spend less time doing activities they enjoy. It is also clear that lower income correlates with unhappiness and that increases in salary lead to only transient happiness due to the phenomenon of "adaptation."

Among other factors, even college education has little to do with happiness but clearly correlates with stress. Having children is the biggest contributor to unhappiness; they lead to constant feelings of stress, sadness, and worry.[12]

In one study, actors were asked to express feelings of happiness and sadness while examined with fMRI.[13] In both states, activation occurred in the frontal lobes, anterior temporal lobes, and the pons. Although the regions were similar for both emotional states, different subregions were activated for each. In a different fMRI study, the mode and tempo of music were manipulated to be perceived as either sad or happy and the former elicited responses in the left orbito- and mid-dorsolateral frontal cortices.[14] Happy voices elicit stronger and different fMRI responses than angry ones.[15] Body postures may also indicate happiness or other emotions.

When observing human body postures, our brain always records 2 things: action and emotion. These states activate visual representation/motion processing and emotional interpretation areas. Both areas are activated simultaneously but differently in men than in women.[16] Men seem to show more reliable activation but in lesser amounts than women.

It seems to me that we neuroradiologists have every reason to be happy, and despite that, only a small group of us are generous with our money and time. Generosity is generally encouraged by the so-called "immediacy bias," better known as a "call to action."

Crises and feelings of uncertainty and worry lead to greater donations.

What better call to action than the lack of scientific evidence of what we do and the ever-decreasing government funding of research? If we do not support our Foundation, these issues will never be solved. If we continue to be as happy as we are now and do not increase our generosity by contributing to our Foundation, our jobs and other sources of happiness will soon disappear.

* Only 7% of ASNR members donated to our Foundation in 2012 (and most were members of the Executive Committee). The total donations for that year were slightly over $300,000, but this included 7 corporate donors and 1 practice group donor. Conversely, more than 60% of the ASNR staff, who

get paid much less than physicians (average salary for radiologists in the United States: $349,000 per year in 2012), donated to the Foundation last year.

REFERENCES

1. Brooks AC. **A nation of givers.** *The American* 2008 http:// www.american.com/archive/2008/march-april-magazine-contents/ a-nation-of-givers. Accessed September 17, 2013
2. Avinun R, Israel S, Shalev I, et al. **AVPR1A variant associated with preschooler's lower altruistic expression.** *PloS One* 2011;6:e25274
3. Zak PJ, Stanton AA, Ahmado S. **Oxytocin increases generosity in humans.** *PloS One* 2007;2:e1128
4. Zak PJ. *The Moral Molecule: The Source of Love and Prosperity.* New York: Dutton Adult; 2012
5. James RN, O'Boyle MW. Charitable estate planning as visualized autobiography: an fMRI Study of its neural correlates. February 6, 2012. Social Science Research Network. http://papers.ssrn.com/sol3/ papers.cfm?abstract_id_2000345. Accessed September 17, 2013
6. Powers R. *Generosity: An Enhancement.* New York: Farrar, Straus, and Giroux; 2009
7. Hibbing JR, Alford J, Lohrenz T, et al. **Generosity is its own reward: the neural basis of representation.** In: *Proceedings of the Annual Meeting of the American Political Science Association,* Toronto, Canada; September 3–6, 2009. http://ssrn.com/abstract_1451309 Accessed on September 17, 2013
8. Davis MA. **Understanding the relationship between mood and creativity: a meta-analysis.** *Organizational Behavior and Human Decision Processes* 2009;108:25–38
9. Graves JA. **The 25 best jobs of 2013.** *US News & World Report Money Careers.* http://money.usnews.com/money/careers/slideshows/the-25-best-jobs-of-2013. Accessed September 17, 2013

10. Physician Compensation Report 2013. Medscape Multispecialty. http://www.medscape.com/features/slideshow/compensation/2013/ public. Accessed September 17, 2013
11. Kahneman D, Deaton A. **High income improves evaluation of life but not emotional well-being.** *Proc Natl Acad Sci U S A* 2010;107:16489–93. www.pnas.org/cgi/doi/10.1073/pnas.1011492107. Accessed September 17, 2013
12. Kahneman D, Krueger AB, Schkade D, et al. **Would you be happier if you were richer? A focusing illusion.** *Science* 2006;312:1908–10
13. Pelletier M, Bouthillier A, Le´vesque J, et al. **Separate neural circuits for primary emotions? Brain activity during self-induced sadness and happiness in professional authors.** *Neuroreport* 2003;14:1111–16
14. Khalfa S, Schon D, Anton JL, et al. **Brain regions involved in the recognition of happiness and sadness in music.** *Neuroreport* 2005;16:1981–84
15. Johnstone T, van Reekum CM, Oakes TR, et al. **The voice of emotion: an fMRI study of neural responses to angry and happy vocal expressions.** *Soc Cogn Affect Neurosci* 2006;1:242–49
16. Kana RK, Travers BG. **Neural substrates of interpreting actions and emotions from body postures.** *Soc Cogn Affect Neurosci* 2012;7: 446–56

THE SOCIAL NETWORK OF LONELINESS

Unlike other countries, we Americans like our heroes to be lonely, and some that come to mind immediately are cowboys, explorers, and scientists patiently working in their laboratories while remaining decontextualized from their surroundings, and perhaps the loneliest of them all: astronauts. Thus, it should come as no surprise that in some of the better recent movies, *Gravity* (Sandra Bullock) and *All Is Lost* (Robert Redford), the main characters find themselves completely alone, and at the end, they are portrayed as heroes. As our social exchange structure changes, Americans are more lonely than ever before, despite our increasing population and our ability to communicate with each other more often and faster.

Although one can argue about the differences and similarities of the following terms: aloneness, isolation, retreat, and seclusion, what I would like to briefly address here is "loneliness," which I take to mean a lack of companionship that may occur even when surrounded by or "connected" to others. Nowadays, our connections are basically electronic and, to many my age or younger, accomplished through Facebook and other "social media."

As of this writing, our main modern social communication tool, Facebook, had 1.31 billion subscribers and 680 million mobile users.[1] Here are some more Facebook statistics that amaze me: 640 billion minutes per month are spent on it, nearly 50% of those 18–34 years of age use it, and it has more than 1 trillion page views per month and 2.7 billion "likes" every day. At the time of this writing, the *American Journal of Neuroradiology* (*AJNR*) had 5461 "likes" and *Radiology* had 28,521 "likes" on their Facebook pages. Thus, it seems that we radiologists are indeed, true Facebook aficionados. For those who like a more concise communication, it will be a relief to learn that Twitter is not doing badly at all. It has nearly 646 million users and hosts nearly 10,000 tweets per second,[2] and just to be fair to Google, I need to mention that its social network (Google_) now has more than 300 million monthly users who upload 1.5 billion pictures every week.[3]

As our electronic social media grow, we seem to get lonelier.

The number of US households tripled between 1940 and 2010, but while in 1940, 90% of them contained families, in 2010 only 66% did.[4] About 27% of households have only 1 person, a number 3 times higher than 50 years ago, and 33% of households now have childless couples.

Is loneliness biologic? Cole et al[5] from the University of California, Los Angeles (UCLA) published, in 2007, an interesting article on this topic. He and his coauthors suggested that changes in genes that are related to inflammation also drive chronic high levels of feelings of isolation and loneliness. This study revealed that the levels of gene expressions may be different depending not only on how many people you know but also related to the number of those you feel close to. Intuitively, a relationship between feeling lonely and one's immune system makes sense to me. The greater the number of close friends one has, the more your immune system must be ready to combat the germs they carry. Conversely, a lesser number of friends may result in a lazier immune system, making your health more fragile; it is well-known that the lonely have precarious health. The way the brain perceives and reacts is also different in the lonely. When examined with fMRI, lonely individuals showed less activation of the ventral striatum, which correlated with a feeling of being less rewarded by social stimuli.[6] Non lonely people showed higher activation of this region, implying that social interactions resulted in a pleasurable event. Lonely individuals also appeared to be more drawn to the distress of others. These studies and others seem to indicate that a lack of perceived pleasure from social interactions is at the core of loneliness.

To avoid loneliness, one must have personal relationships—that is, having a lot of friends on Facebook will not relieve one's feelings of isolation. Conversely, it could be that lonely individuals spend all of their time on Facebook trying to build up a large network of "friends." Moira Burke from the Carnegie Mellon University studied Facebook users and concluded that only those for whom Facebook served as a conduit to establishing direct communications with other individuals leading to friendship seemed to avoid feelings of loneliness.[7] That is, having a large number of friends write on your Facebook wall or communicating with them by terse Twitter-like exchanges will not decrease loneliness. Another study concluded that if one has a lot of friends in real life, one will also have a lot of them on Facebook and be a successful user of it.[8] Simply consuming and broadcasting trivial life events on social media makes one more, not less lonely. The popularity of Facebook may reflect the increasing desire to find oneself among friends (31% in 2010 versus 37% today). Groups of individuals who make and keep friends easier are the Millennial generation (47%), Hispanics (47%), and never-married adults (44%).[9]

To measure loneliness, one of the most popular methods is the UCLA scale. This 20-question scale is easy to use and apparently reproducible. You may find it at http://www.tactileint.com/portfolio/uclalone.html, and when I took the test, I scored a 19, which is the average score for school teachers (I guess I must share with them some frustrations and feelings of isolation perhaps even leading to loneliness). Using this scale, the American Association for Retired Persons (AARP) has found that 35% of adults consider themselves lonely, especially those in poor health, the socially isolated, those with a new residence (less than 1 year), and females; but it also concluded that as we get older overall we feel less and not more lonely.[10] With respect to electronic communications, AARP found that those using e-mail felt that they had fewer deep friendships than before. Not surprisingly, AARP also reported that isolation and loneliness increase a person's risk of death. Loneliness increases circulating cortisol levels that may contribute to brain and cerebrovascular disease and affect sleep patterns that may lead to depression. The common threads between heart and brain vascular disease could be related to the fact that the lonely have an increased incidence of hypertension and smoke more (on average, 15 cigarettes per day).

Is loneliness genetic in nature? Some studies suggest a strong genetic effect, but because loneliness is highly influenced by environmental situations, its expressions vary from childhood to adulthood and from individual to individual. In 1 highly regarded study, loneliness was assessed in 8387 twins.[11] The conclusions of the study were that heritability of differences in feelings of loneliness was 48%, no unique environmental influences were discernible, genetic contributions to loneliness were similar in children and adults, and the heritability of loneliness showed no sex differences.

The authors suggested that individuals are unable to control their loneliness as a response to external stimuli.

Why does loneliness hurt? One hears others saying that "they are so lonely it hurts." Because social exclusion is a type of loneliness, one fMRI study assessed the brains of individuals who were excluded from a specific activity.[12] The results paralleled those from studies of physical pain. That is, the anterior cingulate cortex was less active during exclusion and correlated with self-reported distress. The right ventral prefrontal cortex was active during exclusion, suggesting that it regulates feelings of loneliness by excluding the function of the anterior cingulate cortex. From a developmental standpoint, the same authors believe that the loneliness system "borrowed" its computations from the pain system to prevent its harmful effects.

Perhaps some of the loneliest moments may be experienced within a marriage. Married people are healthier, live longer, and are less lonely but only if their spouses are confidants. This is, again, loneliness is related to the quality of the relationship and not to the relationship per se. One's immediate circle of confidants also extends to one's best friends. Nevertheless, how often you do hear American adults say they have a best friend to confide in? The answer is not often. In 1985, the number of confidants a person had was close to 3; in 2004, it had decreased to 2; and today 25% of Americans claim to have no one to confide in.[13] This seems to be paradoxical when the average American has 634 electronic social ties, but the truth is that most if not all of these ties are superficial and eventually meaningless. When Facebook data are analyzed, it has been found that the largest single group (22%) of "friends" a user has consists of people he or she knew in high school followed by extended family and coworkers. I personally feel that the connection between myself and those I went to high school with is now basically nonexistent, but then, I do not have a Facebook page (I do have a Facebook account that I use to check *AJNR's* page) and do not respond to any Facebook invitations.

What is even worse is that up to 7% of Facebook "friends" are strangers whom the user has not and will never meet.

Since our ever-expanding dependency on electronic communications seems to be making us lonelier, it seems ironic that several sites, such as the Web of Loneliness, offer on-line help via chat rooms and blogs and other types of virtual support groups, many through Facebook and Twitter. Another such site is the UK's Campaign to End Loneliness, which, again, contains a plethora of posts (most are useless) and some pictures of their followers, mostly octogenarians whom I doubt know how to use Twitter or Facebook. These sites state that 5 million older British individuals have, as their sole companion, their television. Of course, many of these sites have a scam-like scent and accept donations via Pay Pal, but some like the UK one are supported by philanthropic foundations.

As in many other situations in our lives, loneliness is multifactorial.

Our pursuit of space and individualism (an idea ingrained in American culture) and the desire to be alone drove city populations into the suburbs and beyond. So do as country music legend Willie Nelson says: "Mamas, don't let your babies grow up to be cowboys, they'll never stay home and they're always alone, make'em be doctors and lawyers and such. . . ." Moreover, I should add: keep them away from Facebook and Twitter.

REFERENCES

1. STATISTIC BRAIN. Facebook Statistics. http://www.statisticbrain. com/facebook-statistics. Accessed March 19, 2014
2. STATISTIC BRAIN. Twitter Statistics. http://www.statisticbrain. com/twitter-statistics. Accessed March 19, 2014
3. Barr A. **Google's social network sees 58% jump in users.** *USA TODAY.* October 29, 2013. http://www.usatoday.com/story/tech/2013/10/29/google-plus/3296017. Accessed November 11, 2013
4. acobsen LA, Mather M, Dupuis B. Population Reference Bureau. Household change in the United States. http://www.prb.org/ Publications/Reports/2012/us-household-change.aspx. Accessed November 11, 2013
5. Cole SW, Hawkely LC, Arevalo JM, et al. **Social regulation of gene expression in human leucocytes.** *Genome Biol* 2007;8:R189
6. Cacioppo JT, Norris CJ, Decety J, et al. **In the eye of the beholder: individual differences in perceived social isolation predict regional brain activation to social stimuli.** *J Cogn Neurosci* 2009;21:83–92
7. Burke M, Kraut R, Marlow C. **Social capital on Facebook: differentiating uses and users.** *CHI* 2011. May: 7–12, 2011. http://citeseerx. ist.psu.edu/viewdoc/download?doi_10.1.1.227.6644&rep_rep1& type_pdf. Accessed May 5, 2014
8. Quercia D, Lambiotte R, Stillwell D, et al. **The personality of popular Facebook users.** http://dl.acm.org/citation.cfm?id_2145346. Accessed April 24, 2014

9. BarnaGroup. How the last decade changed American life. https:// www.barna.org/barna-update/culture/624-how-the-last-decadechanged-american-life#.UoEdduKO51N. Accessed April 24, 2014
10. Anderson G. **Loneliness among older adults: a national survey of adults 45_.** http://www.aarp.org/personal-growth/transitions/info-09-2010/loneliness_2010.html. Accessed November 11, 2013
11. Boomsma DI, Willemsen G, Dolan CV, et al. **Genetic and environmental contributions to loneliness in adults: the Netherlands twin registry study.** *Behav Genet* 2005;35:745–52
12. Eisenberger NI, Lieberman MD, Williams KD. **Does rejection hurt? An fMRI study of social exclusion.** *Science* 2003;302:290–92
13. Kornblum J. **Study: 25% of Americans have no one to confide in.** *USA TODAY*. June 22, 2006. http://usatoday30.usatoday.com/news/nation/2006-06-22-friendship_x.htm. Accessed March 19, 2014

Famous people:

PIGEONS AND MRI: TESLA VIGNETTES

Nikola Tesla's name is familiar to neuroradiologists, but few of us know why our profession and understanding of the brain are better because of him. His achievements were not universally recognized until 1960 (17 years after his death) when the *Conference Generale des Poids et Mesures* decided that the unit for measuring the magnetic field (B) should be called the "tesla." The strength of the magnetic field of the earth at the equator is 31 microteslas, and it is worth remembering that magnets found even in small speakers are as powerful as those found in our MR imaging units (1–2.4T).[1] His research also involved superconducting magnets cooled to a few degrees above absolute zero.

Although Tesla was constantly on the verge of becoming rich, he died poor and alone at 86 years of age in a room at the Hotel New Yorker (still at 481 Eighth Avenue). After his death, unknown individuals and/or government forces removed his last inventions and papers from his apartment because they were thought to contain information regarding the "Death Ray," in which the military was interested.[2] This ray, presumably a particle beam, could repel armies and bring down airplanes. The list of Tesla's inventions is long and incredible in its breadth. During his life, Mr. Tesla struggled for recognition, and it mostly eluded him, the Nobel Prize being one example. In this Perspectives, I mention some colorful aspects of Mr. Tesla's life rather than recounting his incredible accomplishments.

Tesla and Edison

Tesla came to America after living in France for 2 years, and Mr. Edison hired him to work at the Edison Machine Works in New York City. Edison's electric power generators producing direct current (DC) worked well only when electricity requirements were small. The power and output of DC are relatively weak, making its transmission over long distances impractical. Tesla solved this issue by perfecting alternating current (AC). Transformers decrease or increase the power of AC as needed, so if long-distance transmission is required, power is amplified, making it more efficient (with DC current, a generator every 3–4 km is needed).

When Edison refused to pay him for his inventions, Tesla left and later sold them to Westinghouse. The War of Currents erupted, and Edison tried to instill fear in AC users by calling it extremely dangerous. He went as far as paying children to steal dogs, electrocute them with AC, then scatter their bodies together with flyers alerting the public to the dangers of AC. When killing dogs no longer shocked the public, he killed larger animals (sheep, cows, horses), eventually leading to the electrocution of Topsy the elephant (for a video of this see: http://tinyurl. com/mumu8mq and, for a good description of the process, I suggest reading Jean Echenoz's fictional Tesla biography *Des Eclairs*[3]).

Additionally, Edison promoted the first successful electrocution of a prisoner by using AC (DC was tried but was not powerful enough to kill a human being) to showcase the dangers of this type of electricity. The subsequent legal battles that erupted nearly caused Edison and Westinghouse to go broke and forced

Tesla to forfeit royalties from his patents owned by the latter. When Edison died, Tesla wrote this bitter obituary: "He had no hobby, cared for no sort of amusement of any kind and lived in utter disregard of the most elementary rules of hygiene. . . .His method was inefficient in the extreme, for an immense ground had to be covered to get anything at all unless blind chance intervened and, at first, I was almost a sorry witness of his doings, knowing that just a little theory and calculation would have saved him 90% of the labor. But he had a veritable contempt for book learning and mathematical knowledge, trusting himself entirely to his inventor's instinct and practical American sense."[4] Just before dying, Edison acknowledged that ignoring Tesla's AC patent had been his biggest mistake.

Tesla and Roentgen

Before x-rays were officially named, Tesla investigated them by using single-terminal vacuum tubes (conventional ones use 2 terminals).

We know that high-energy electrons emitted by a cathode hit the special material (tungsten, molybdenum) of the anode, "braking" them and secondarily emitting a very small percentage of high-energy x-rays. In Tesla's tube, no target existed.

Energy left the electrons encountering a high-field electrical environment resulting from the oscillations of AC, and as they collided with the glass encasement, x-rays were generated. His experiment also worked well by using Geissler tubes that were filled with substances such as inert gasses (these were the forbearers of fluorescent light and the electron microscope). While in New York, he produced images of the bones in his hands and sent them to Roentgen, who ignored them. Tesla also claimed that his design produced x-rays much more powerful than Roentgen's.

Because Tesla never published his findings and his research notes were lost during a fire of suspicious nature, he never received credit for the discovery of x-rays. Fortunately, he also became aware of Roentgen's health issues induced by radiation exposure and avoided them himself.

Tesla and Marconi

Although Guglielmo Marconi is credited with having invented the telegraph and received the Nobel Prize for the radio, Tesla discovered both years before Signore Marconi did. Tesla discovered that by using his coils, radio signals could be transmitted over great distances as long as the receiving coil was tuned to the resonant frequency of the transmitting one (sound familiar?). The receiving coil magnifies signals via resonance. Just before Tesla could demonstrate that his invention was able to transmit signal as far as 50 miles, his laboratory suspiciously burned down, causing him to lose his instruments and documents (note that Tesla demonstrated transmission at shorter distances in St. Louis 2 years before Marconi showed his invention). About the same time as the aforementioned tragedy, Marconi developed a 2-way transmitter whose signals were too weak to cross even a small pond. He solved the problem by using Tesla coils. Marconi claimed ignorance about Tesla's coils when applying for a US patent, and the granting of patents to both inventors was delayed due to arguments on both sides claiming property rights. Aided by rich investors, Marconi's Wireless Telegraph Company thrived in the stock markets, and soon Andrew Carnegie and Thomas Edison became its 2 most important American investors. Shortly thereafter, Marconi amazed the world by transmitting signals wirelessly across the Atlantic Ocean. Because Marconi was using several of Tesla's patents to accomplish this, Nikola was not worried; but he should have been because 4 years later, the United States awarded the patents for the telegraph and radio to Marconi under political pressure from Carnegie and Edison. When Marconi received the Nobel Prize, Tesla was furious and sued Marconi for stealing his patents. Later, the Marconi Company sued the US government because the armed forces had used its patents for communications during World War I without permission or payment. The US Supreme Court eventually ruled that these patents belonged to Tesla (now dead and childless), thus avoiding any payments owed by the government to Marconi's company.[5] Tesla predicted that all of us would carry small, wireless telephones in our pockets; something Marconi did not.

Tesla and Twain

Exactly where Tesla and Samuel Clemens met is not clear; it could have been at the Player's Club (a bar in Manhattan) or in the laboratory. Although Tesla was familiar with Twain's writings, it was not until after the discovery of AC that Twain noticed Nikola.

Both men shared friends in high society, including the Johnsons, Kipling, Roosevelt, and Muir. Tesla invited Twain to his laboratory, where the famous writer partook in some electrical experiments that reportedly filled him with vigor and vitality. While Twain was spending time in Austria, he heard about Tesla's experiments on destructive terror (the Death Ray) and wrote urging him to use these to make war impossible in the future by making it available to all (an idea akin to "assured mutual destruction").

Tesla and the FBI

While living in Colorado Springs, Tesla started developing the idea for a particle beam that could be used as a weapon, and though his idea never materialized, it was described in what is known as the "Tesla Papers." Immediately after his death, unknown persons raided his apartment in the New Yorker Hotel, stealing documents for fear that they would fall into Soviet hands.

Two days after his death (when he was found by a maid), the FBI confiscated all that was left. The FBI appointed Dr John Trump of the National Research Committee to look into the documents, and he concluded that they were mostly speculative in nature.

After World War II, interest in them was revived and the heavily funded "Project Nick" was started in Dayton, Ohio, only to be dropped later. Interest in a beam weapon waxed and waned until the late 1970s, when construction of a large beam weapon by the Soviets came to light. As a response, in the early 1980s, President Ronald Reagan proposed the Strategic Defense Initiative.[6] All attempts to build a Death Ray have failed, and many think that answers to the problems encountered were addressed in the missing "Tesla Papers " (some think that the US Government has them and is hiding them).

Tesla and Birds

No one knows why Tesla was interested in pigeons. This interest is nicely portrayed in *The Invention of Everything Else* by Samantha Hunt.[7] As an old man, Tesla walked every day to Bryant Park (located behind the New York Public Library between 6th Avenue and West 42nd Street, a scene again found in Paul Auster's *Moon Palace*[8]). At that time, pigeons were considered unmeritorious, and perhaps Tesla felt similarly about himself. On the night that he was awarded the Edison Medal (how ironic!), he suddenly disappeared from the banquet only to be found in Bryant Park covered by pigeons from head to evening pumps. Tesla said he considered pigeons to be his "sincere friends." He took sick ones into his apartment and caused cleaning crews to complain of dirt.

Just days before his death, he became particularly attached to one and was able to recognize this particular bird and fed it every day (white wings with a touch of gray in their tips, photographs available at: http://www.teslauniverse.com/nikola-tesla-timeline-1922-tesla-pigeon-dies). When the pigeon became sick, Tesla took it with him to his apartment, but attempts to cure the bird failed and it died. Tesla died only a few days after the pigeon, it is said of a broken heart (this may be true because he died of heart failure); he previously stated that he had loved her as a man loves a woman (how sad is that?). Much has been made about the symbolism of the pigeon, comparing it with the dove in religion and its meaning in Tesla's life. Curiously, Tesla's favorite meal was squab.

Tesla, like Einstein, was a generalist, and like Edison, he was self-taught. His thoughts extended into many arenas of human enterprise without dwelling on details of how to accomplish them.

Because he almost never published in scientific journals, many of his ideas are now lost. Some of his projects sounded like science fiction but are now reality; others are still within the realm of the impossible but are being reconsidered. Our knowledge has expanded so much that extrapolating what we now know into the world where Tesla lived is simply not possible, so it is hard for us to grasp his achievements. However, it is thanks to them that MR imaging and modern neuroimaging are possible.

Bonus

Two wonderful stories about Edison and Topsy, and Tesla and his pigeons can be found in *Love in Infant Monkeys* by Lydia Millet.[9]

REFERENCES

1. Tesla (unit). Wikipedia. https://en.wikipedia.org/wiki/Tesla_(unit)

2. Lost Tesla Papers about "Death Rays." Tesla Memorial Society of New York. http://www.teslasociety.com/deathray.htm. Accessed on June 5, 2013
3. Echenoz J. *Des Eclairs*. Paris: Les Editions de Minuit; 2010
4. Nikola Tesla. Wikipedia. http://en.wikipedia.org/wiki/Nikola_Tesla. Accessed on June 5, 2013
5. http://www.pbs.org/tesla/ll/ll_whoradio.html. Accessed on June 5, 2013
6. Tesla, Life and Legacy. The Missing Papers. PBS. http://www. pbs.org/tesla/ll/ll_mispapers.html. Accessed on June 5, 2013
7. Hunt S. *The Invention of Everything Else*. Boston: Mariner Books; 2008 **826** Editorials May 2014 www.ajnr.org
8. Auster P. *Moon Palace*. New York: Viking Press; 1989
9. Millet L. *Love in Infant Monkeys*. New York: Soft Skull Press; 2009

APPLES

Dip the apple in the brew, let the sleeping death seep through.
The Evil Queen in Snow White and the Seven Dwarfs

Why Steve Jobs chose an apple as the symbol for his cool company is not certain: Was it working at an apple farm when he was young or admiring the Beatles, Isaac Newton, and Alan Turing? The apple he chose to represent his line of computers was a Macintosh. No doubt apples are attractive: wide shoulders leading to feminine rounded sides ending in a wide stem bowl, superiorly and inferiorly in a narrow calyx, all wrapped in a colorful skin freckled with lenticels, and they are also good for you—2 (or 3) apples a day keep the neurologist away. Animal studies prove that antioxidants, acetylcholine and quercetin, occurring in apples are good for the aging brain and may even help prevent Alzheimer disease.[1] Apple juice is said to improve the mood of patients with

Alzheimer disease, and some compounds in apples may help fight the disease.[2] Apples also help the circulatory system by decreasing serum low-attenuation lipoproteins (bad cholesterol). Hence, there must be some truth to the old adage "An apple a day keeps the doctor away," which comes from the even older English version, "An apple before going to bed keeps the doctor from winning his bread."[3] However, apples are also associated with evil behavior and forbidden love.

On June 7, 1954, Alan Turing was 42 years of age, fat and flabby with enlarged breasts, when he carefully injected cyanide into an apple—a red one, his favorite—and started eating it. The next day he was found dead with half of that apple beside him.[4]

Before his suicide, he was working on mathematic representations of biologic processes now considered the fore bearers of the Chaos Theory and the Singularity Event. His purpose was to discover how computers might merge with human processes. That same year he had imagined a computer that could play chess. For nearly all of the previous decade he had been designing the Automatic Computing Engine while working at the National Physical Laboratory in London.[5] His article "Computing Machinery and Intelligence" published in 1950 is said to be the most quoted paper in modern philosophic literature.[6] During this time, he established a homosexual relationship, and one of his partner's friends broke into his house at night to steal. Turing reported this incident to the police, confessed his homosexuality, and was charged with gross indecency. Punishment: chemical castration (which explains his physical condition and psyche at the time of his death).

Chemical castration induces depression, anxiety, and decreased cognitive abilities and verbal memory.[7] Androgen blockade also leads to increased levels of cerebral _-amyloid, which is associated with Alzheimer disease. Conversely, estrogen administration to postmenopausal women leads to lower levels of _-amyloid and improved cognition. Although depression is a major risk factor for suicide, suicidal individuals have other more pronounced brain chemical abnormalities than those who are "only" depressed. Among those who commit suicide, levels of corticotropin- releasing hormone in the red nuclei and several parts of the frontal lobes are increased. Other hormones are also affected: vasopressin is high in some brain regions and low in others. Less well-known hormones such as gastric-releasing peptide are high, and neuromedin B is low. Depression may also exert epigenetic influences by increasing the methylation of a gene that regulates the expression of DNA methyltransferase, leading to abnormal function of the frontopolar cortex.[8] Thus, although suicidal tendencies are probably not genetic, environmental factors can drive genetic changes that lead to them.

When he was 18 years of age, Turing fell in love with a fellow student who later died of tuberculosis acquired from drinking unpasteurized milk. This death shattered all his religious convictions and strengthened his interest in mathematics. As a teenager, he was sent to study in Dorset. He was said to have been disorganized, disheveled, and perhaps not too clean. He had a strange voice, high-pitched and halting, that did not change with adolescence and separated him from other boys. As a young boy, one of his favorite stories was Snow White, particularly the part when the evil queen cooks the poisoned apples. The queen tricks Snow White into eating one by making her believe that it is a magic wishing apple.

Homosexual men show brain differences compared with others.

The volume of their anterior commissure is larger than that of heterosexual males and females. The suprachiasmatic nuclei are also larger in gay men.[9] No one knows when these differences are established. A prenatal surge in luteinizing hormone and a decrease in testosterone are thought to influence sexual orientation.

[10] The volumes of the cerebral hemispheres are nearly identical in homosexual men and heterosexual women but different from those of heterosexual men and lesbians. Functional connections as inferred from PET studies follow these same characteristics.

[11] Different types of auditory and masticatory patterns and increased left-handedness are found in gay men. Today, in our Western world, punishing someone for his or her sexual preference sounds extreme, but it is to be remembered that homosexuality is still punishable by death in 7 countries and illegal in nearly 40% of all countries.

Most of Turing's fame came from his work as a cryptographer during World War II. Initially he was able to solve the "indicator system" used in cypher texting, which showed the initial position of the rotors in a coding machine before messages were transmitted.

On the same day that he solved the indicator issue, he also conceived of a technique that would break the secrecy of the Nazis' famed Enigma code (used to transmit instructions to U-boats). Because the Enigma was capable of 4 _ 10^{26} combinations, just having the chutzpah to imagine a method to break its codes is mind-blowing.[12] These techniques led to the creation of the first programmable computers (the Heath Robinson and the Colossus), though he was not directly involved in building these.

Before working for Naval intelligence, he spent 2 years in the United States at the Institute for Advanced Study at Princeton, where he obtained his PhD. The topics of his dissertation were ordinal logic and relative computing, which helped "Turing machines" solve problems not hitherto possible. Turing machines, also known as "universal machines," were intended to solve any- thing that is computable.[13] These machines were capable of manipulating symbols found on a strip of paper (or other materials) separated by other symbols (colons, semicolons) or cells and were the ancestors of modern central processing units. On a strip of paper, a black square may have signified zero, while a white one meant 1. Thus, the binary code was put to practical use, and Turing used numbers to represent other numbers, creating modern computing. Because computers can compute forever without knowing when to stop, Turing solved this issue by creating an algorithm called the "halting problem," which represents the first decision-making program.

In the late 1940s, Turing began tinkering with artificial intelligence (AI) (he is now called "the father of AI"). During this time, he developed the "Turing test," intended to determine a machine's ability to exhibit intelligent and humanlike behavior. For this, he asked a human judge to engage in conversation with another human and a machine, not knowing which one was which. If the judge was unable to differentiate between these, the machine passed the test and was considered intelligent. It was not until 1966 that a computer program passed this test. The reverse of this test type is called CAPTCHA, and it ensures that a response is human.[14] All of us are familiar with this test because it is commonly found when registering for on-line services as a series of letters and/or numerals that are distorted and crowded, making segmentation difficult. These must be retyped correctly by a human user:

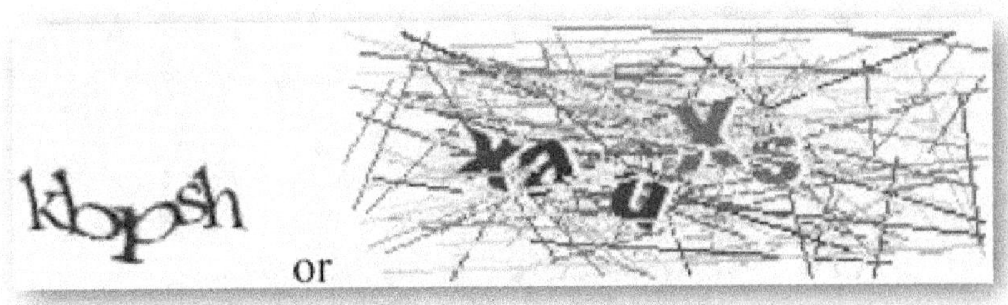

The apple has much more to do with modern neuroradiology than just Turing and his machines. Electrical and Musical Instruments (EMI) is a British consortium that developed radar and stereo sound recording.[15] Perhaps, the best known division of EMI is Abbey Road Studios, which opened in 1931, the year EMI was formed. Before Sir George Martin brought the Beatles to the studios, classical luminaries such as Elgar, Toscanini, and Klemperer recorded there. In 1967, the Beatles started Apple Corps (comprising a group of companies) and issued their first Apple recording, "Magical Mystery Tour." Although EMI considered itself a "serious company" and viewed Abbey Road Studios as its black sheep, there is no denying that together with Apple Records, they brought it tremendous wealth. It is with this income that 4 years after Turing's death, Sir Godfrey Hounsfield built the first fully transistorized computer (EMIDEC 1100) by using many of the concepts invented by Turing. The EMIDEC 1100 manipulated information encoded in punched cards or tapes similar to the Turing machine. In 1967, after EMI gave Hounsfield "time to think about some research which may be fruitful," he started exploring the idea of pattern recognition with computers.[16] One year later, EMI patented Hounsfield's idea for body scanning by using x-rays, and in 1971, the first brain CT study was done at the Atkinson Morley Hospital in London. It is said that thanks to the sales of 200 million vinyl Beatles singles, EMI was able to fund the 4 years that it took Hounsfield to realize his idea.[17]

Nowadays, Apples (computers, not fruit) are said to be addictive.

In 2011, the British Broadcasting Company screened a documentary, and in it, a devotee of Apple computers underwent an fMRI study that showed that the brand stimulated the same brain areas as religious faith does.[18] "Applephilia" can become a disease, and some individuals claim to think about Apple 24 hours a day.[19]

In 1966, the first Turing Award was given and is now considered the Nobel Prize of computing. The award comes with US $250,000 and is sponsored by Intel and Google. No person related to the field of medical imaging has received it.

REFERENCES

1. Heo HJ, Kim DO, Choi SJ, et al. **Apple phenolics protect in vitro oxidative stress-induced neuronal cell death.** *J Food Sci* 2004;69:S357–60
2. http://www.sciencedaily.com/releases/2004/11/041116215006.htm. Accessed April 2, 2012
3. Seabrook J. **Crunch**. *The New Yorker.* November 21, 2011:54–64
4. Hodges A. **The Alan Turing homepage.** http://www.turing.org.uk/ turing. Accessed April 2, 2012
5. Alan Turing. Wikipedia. http://en.wikipedia.org/wiki/Alan_Turing. Accessed April 2, 2012
6. Alan Turing. http://plato.stanford.edu/entries/turing. Accessed April 2, 2012
7. Almeida OP, Waterreus A, Spry H, et al. **One year follow-up of the association between chemical castration, sex hormones, beta-amyloid, memory and depression in men.** *Psychoneuroendocrinology* 2004; 29:1071–81
8. Poulter MO, Du L, Weaver CG, et al. **GABAA receptor promoter hypermethylation in suicide brain: implications for the involvement of epigenetic processes.** *Biol Psychiatry* 2008;64:645–52
9. Allen LS, Gorski RA. **Sexual orientation and the size of the anterior commissure in the human brain.** *Proc Natl Acad Sci U S A* 1992;89:7199–202
10. Balthazart J. **Minireview: hormones and human sexual orientation.** *Endocrinology* 2011;152:2937–47

11. Savic I, Lindstrom P. **PET and MRI show differences in cerebral asymmetry and functional connectivity between homo- and heterosexual subjects.** *Proc Natl Acad Sci U S A* 2008;105:9403–08
12. The Enigma machine. http://www.murky.org/blg/2004/09/theenigma-machine. Accessed April 2, 2012
13. Kamvysselis M. **Universal Turing machine.** http://web.mit.edu/ manoli/turing/www/turing.html. Accessed April 2, 2012
14. ReCAPTCHA. **Telling humans and computers apart automatically.** http://www.google.com/recaptcha/captcha. Accessed April 2, 2012
15. EMI. Wikipedia. http://en.wikipedia.org/wiki/EMI. Accessed April 2, 2012
16. The Nobel Prize in physiology and medicine 1979. Allan M. Cormack, Godfrey N. Hounsfield. http://www.nobelprize.org/nobel_prizes/medicine/ laureates/1979/hounsfield-lecture.html. Accessed April 2, 2012
17. The Beatles greatest gift…is to science. http://www.whittington. nhs.uk/default.asp?c_2804. Accessed April 2, 2012
18. Popken B. **MRI shows Apple stimulates fan's brain like religion.** THE CONSUMERIST. http://consumerist.com/2011/05/mri-showsapple-stimulates-fans-brains-like-religion.html. Accessed April 2, 2012
19. Brooks A. **Shareholder meeting reveals little insight.** http://www.worldofapple.com/?s_shareholder meeting reveals little insight. Accessed April 2, 2012

N.B. For a wonderful fictional history of Turing, I recommend *A Madman Dreams of Turing Machines* (New York: Anchor Books, 2006) by Janna Levin, a Professor of Physics and Astronomy at Columbia University. Also The office of Janna Levin at http://www.jannalevin.com.

TINTIN AND COLLEAGUES GO TO THE DOCTOR

I will go ahead and admit it: I ama Tintin fan, albeit a late one as I only started reading his adventures about 10 years ago.

Because the new Steven Spielberg movie will open soon, I relooked at the comic books with special attention to items that may interest neuroradiologists. *The Adventures of Tintin* (Casterman) is a series of 23 books created by the Belgian artist Herge´ (Georges Remi; his initials, backwards and pronounced in French sound like Her-ge). The first book appeared in 1929 and the latest in 1976, 7 years before Herge´ died. Over 350 million copies translated into 80 different languages have been sold to date.[1]

At the start of his adventures, Tintin is said to be about 14–15 years old, but his height is about that of a 7- or 8-yearold American boy today. Curiously, he does not appear to have aged during his long journalistic/investigative career (some Tintinologists disagree and calculate that he aged 3 years during the entire series of books[2]). A normal person would be at least 65 years old by the time *Tintin and the Picaros* (Casterman, 1976) came out and certainly ready to retire, yet he remained youthful and never even shaved! Dr Antoine Cyr, a Professor of Medicine at the University of Sherbrooke in Canada, and his 2 young sons appear to have found a plausible explanation for Tintin's youthfulness. In the 23 books, Tintin loses consciousness 50 times, most often related to trauma.[3]

These count as grade 3 concussions, which are medically characterized by loss of consciousness and amnesia longer than 5 minutes and 24 hours, respectively. It is not certain how many grade 1 or 2 concussions he had. Grade 1 concussions imply no loss of consciousness and amnesia lasting less than 30 minutes.

Grade 2 concussions result in loss of consciousness lasting less than 5 minutes and amnesia between 30 minutes and 24 hours.[4] Not surprisingly, most of Tintin's concussions were the result of being hit in the head with a blunt object, mostly a club, though once he was hit with a camel femur (*The Crab with the Golden Claws,* Casterman, 1941). Cyr et al[3] concluded that Tintin suffered from growth hormone deficiency and hypogonatropic hypogonadism from repeated head trauma.

This diagnosis nicely explains his lack of growth, eternal youth, lack of facial hair, and a lack of libido (Tintin never develops an interest in sex or falls in love throughout the series).

It is curious to note that these characteristics are the ones that endear him to males and females of any age or nationality.

Posttraumatic injury to the pituitary stalk may occur at the time of birth or later in life and result in acquired growth hormone deficiency.[5] Trauma does not have to be severe, and this condition is known to also occur after only mild injuries.

Polyuria (which Tintin does not have) can be absent and is probably due to collateral circulation and/or re-establishment of the hypophyseal portal system. Overall, growth hormone deficiency is the most common pituitary defect after traumatic brain injury.[6] The MR imaging findings of this condition are well known to neuroradiologists: absent stalk, translocated posterior pituitary lobe, and diminished size of the adenohypophysis.

The second most important character in the *Adventures of Tintin* is Captain Haddock who is introduced in the ninth book (*The Crab with the Golden Claws*). From a medical standpoint, Archibald Haddock is as interesting as Tintin. In his first 3 books, he is a weak character and an alcoholic who prefers rum (as I imagine any good sailor does) and whisky.

The worst of his alcoholism shows up in *The Secret of the Unicorn* (Casterman, 1943) where he is in constant need of a drink, is emotionally erratic, confabulates, and lacks any insight into his condition (called "anosognosia"). These features are all compatible with Korsakoff encephalopathy. This syndrome is due to a chronic deficiency of thiamine (vitamin B^1) and results in altered memory, vision changes, and hallucinations, all of which Captain Haddock certainly displays in the *The Secret of the Unicorn*.

There are no specific imaging findings for Korsakoff encephalopathy, though many patients show atrophy of their mamillary bodies. This finding could be the sequela of Wernicke acute encephalopathy, another disease mostly of alcoholics that is also due to thiamine deficiency. MR imaging findings of Wernicke encephalopathy include hemorrhagic and/or enhancing mamillary bodies and increased T2 signal intensity along the walls of the third ventricle and periaqueductal gray matter. Another personality trait of Haddock is his continued and unexpected use of curses and insults (though never profanity). Men, and typically sailors, use this type of language more than women (exception: females in sororities).

Expletives are generally used to convey anger. Haddock certainly has an angry temperament triggered by minimal provocation.

Angry individuals have an increased incidence of hypertension and cardiovascular disease. Moreover, anger leads to carotid artery atherosclerosis, lending some truth to family members not uncommonly reporting that the patient "got so angry, he had a stroke."[7]

Captain Haddock is particularly and constantly annoyed at another character, Professor Calculus. The latter is half-deaf and constantly misunderstands what people are saying, which drives Haddock out of his mind. To make things worse, the Professor nearly always responds to Haddock with exaggerated anger. To be fair, as the series goes on, Captain Haddock tapers his drinking, becomes somewhat heroic, enjoys a social life (rumors of an affair with La Castafiore appear in *The Castafiore Emerald* (Casterman, 1963), thus becoming a more likable chap. This miraculous change stems, in no small part, from his coming into a fortune when he finds a treasure hidden by an ancestor in *Red Rackham's Treasure* (Casterman, 1944)

Some of the funniest characters in the series are a pair of detectives who are identical twins and have names that are nearly identical in written and phonetic form in most languages.

Thompson and Thomson (Dupond and Dupont in French) can be distinguished only by the shape of their mustaches.

They clearly manifest echolalia (the automatic repetition of another's vocalizations) and echopraxia (the automatic repetition of movements made by another person). Echolalia has been documented in autistic twins, but Thompson and Thomson do not appear to be autistic. Patients with Alzheimer disease may also demonstrate echolalia, and because the detectives are quite forgetful, one wonders if they have some form of familial dementia.

Ms Bianca Castafiore is a mature opera singer who is ini- tially despised by Captain Haddock. She is often foolish and absent-minded (Alzheimer disease) repetitively singing the "Jewel Song" from Gounod's *Faust*. Is this a form a palilalia?

"Palilalia" is the immediate repetition of one's own words. In younger children, it is probably normal and forms part of the learning process. La Castafiore probably does not have palilalia as patients affected by it commonly stutter, something she does not. Despite being an opera diva, she suffers from abnormal phonemic awareness and is unable to distinguish words that rhyme (the jokes related to rhyming—originally written in French—lose much of their bite when translated). Of the lesser characters, Nelson the butler is perhaps the one who shows up most. He does not show any physical or psychological abnormalities, but maybe we do not get to know him well enough to detect any abnormality.

Socially, Herge´ committed some faux pas. Take his *Tintin in the Congo* (*Le Petit Vingtie`me*, 1931), where Africans are portrayed as primitive and the overall attitude of the book is paternalistic. Conceivably, this is just a reflection of the spirit of that time, but it has lead to multiple revisions of the book and to it being the last one published in English (it was banned in many countries). Animal cruelty is omnipresent, and stereotyping of individuals (Jews, Native Americans) is also common throughout the series.

Tintin is certainly not the only comic book character to suffer repeated head trauma. A study of traumatic brain injuries has been done in another French-language comic book character: Asterix.[8] This series was created by Rene Goscinny and started in 1959. Asterix is a Gaul warrior resisting the Roman invasion of Gaul in about 50 BC. More than 700 traumatic brain injuries—mostly to males—occur throughout the 34 books. More than 50% of the injuries are moderate with Glasgow Coma Scale scores of 9–12. Thirteen characters show signs of decerebrate posturing. As expected, it is the Romans who suffered the most head injuries, and the most severe ones happened when helmets were not being used. Despite the large number of injuries, it appears that no character suffered long term sequelae.

Digging into the psyches of beloved children's characters may elicit a flurry of complaints and controversies. This is what happened when a group of researchers from Halifax, Nova Scotia attempted to explain the behavior of Winnie the Pooh.[9] The famous bear is now believed to have attention deficit/hyperactivity disorder (ADHD) of the inattentive subtype.

Comorbidity includes impulsivity, cognitive impairment, and finally obsessive fixation (to honey) which leads to . . . obesity. There are bumps on his head suggesting child (or bear) abuse. These modern neurodevelopmentalists suggest that Pooh needs medication to be fitter and more functional.

The other characters do not escape being medically assessed by the authors of the article. Piglet is given a diagnosis of generalized anxiety disorder; Eeyore has a dysthymic disorder; Owl, a reading disorder; and Tigger also has ADHD.

Tintin, Asterix, and Pooh do not go to the doctor in any of the books, but at least in Tintin's case, doctors appear in the comic books. Dr Muller shows up in 3 books (*The Black Island* Casterman, 1938; *The Land of Black Gold*, Casterman, 1950; and *The Red Sea Sharks*, Casterman, 1958). His background and specialty are never given (I am not even sure that he is a medical doctor). Dr Krollspell appears in only 1 installment (*Flight 714*, Casterman, 1968) and is the head of a psychiatric clinic. Later he loses his memory when kidnapped by aliens who give him some undefined "treatment." It is hinted that both of these nefarious characters are ex-Nazis. The last is Dr Patella, who, in accordance with his name, is an osteopath. He appears in 2 books (*Destination Moon*, Casterman, 1953; and *Explorers on the Moon*, Casterman, 1954), where he treats an unconscious Haddock when arriving back to earth.

Should we, and particularly our children, be allowed to read about this bunch of sick characters? Would an older, bearded, libidinous Tintin make more sense? Or, a cowardly Asterix who never fights? Would a leaner more efficient Pooh be a better character? Some think that these characters should go to the doctor; I like them just as they are and prefer not to know what their MR images would show.

References

1. Wikipedia. **The adventures of Tintin.** www.en.wikipedia.org/wiki/The_Adventures_of_Tintin. Accessed on July 15, 2011
2. Tout savoir sur Tintin. www.ydeb.free.fr/Tintin_fichiers/tintin/tintin.htm. Accessed on July 15, 2011

3. Cyr A, Cyr LO, Cyr C. **Acquired growth hormone deficiency with hyponodatropic hypogonadism in a subject with repeated head trauma, or Tintin goes to the neurologist.** *CMAJ* 2004;171:1433–34
4. Cantu RC. **Posttraumatic retrograde and anterograde amnesia: pathophysiology and implications in grading and safe return to play.** *J Athl Train* 2001;36:244–48
5. Yamanaka C, Momoi T, Fujisawa I, et al. **Acquired growth hormone deficiency due to pituitary stalk transection after head trauma in childhood.** *Eur J Pediatr* 1993;152:99–101
6. Popovic V. **GH deficiency as the most common pituitary defect after TBI: clinical implications.** *Pituitary* 2005;8:239–43
7. Bleil ME, McCfferty JM, Muldoon MF, et al. **Anger-related personality traits and carotid artery atherosclerosis in untreated hypertensive men.** *Psychosom Med* 2004;66:633–39
8. Kamp MA, Slotty P, Sarikaya-Weiwert S, et al. **Traumatic brain injuries in illustrated literature: experience of over 700 head injuries in the Asterix comic books.** *Acta Neurochir (Wien)* 2011;153:1351–55, discussion 1355. Epub 2011 Apr 7
9. Shea SE, Gordon K, Hawkins A, et al. **Pathology in the Hundred Acre Wood: a neurodevelopmental perspective on A. A. Milne**. *CMAJ* 2000;163:1557–59

RECOGNIZING FAMOUS FACES

> *The charm of fame is so great that we like every object to which it is attached, even death.*
>
> *—B. Pascal, French mathematician, physicist, and philosopher.*

Part 1: Some Anecdotes

Johnny Cash. Upon boarding an airplane, I turned around, and there he was—the "Man in Black" (truly dressed in black and carrying a black guitar case). I asked for his autograph and, in the absence of a scrap of paper, had him sign my boarding pass only to have that part immediately taken away by the airline clerk. Seeing what was happening, he again signed the little part of the boarding pass I got back (I still have it, and I bet the airline clerk also has hers).

Mick Jagger. While at dinner at a fancy restaurant in Paris, we noticed that the table closest to us was being prepared again. How could that be when there was only 1 service per table? Well, Sir Mick Jagger walked in accompanied by a beautiful Hispanic-looking woman. I have to report that he was simply dressed, behaved very well, and ate very healthy (no wine, just sparking water and fish, no dessert).

Steve Winwood. As I was having lunch one Saturday, multi–Grammy Award winner Steve Winwood walked in and sat beside me. I said, "Hello, Steve," to which he answered by asking me if I had tickets for his concert that night. I lied and said yes. I really do not even like his music.

I have also sat beside Michael Jordan and James Taylor at a local hotel near where I live. Flying back from a meeting, I sat beside the jazz singer Neenah Freelon, who is really as beautiful as she looks on her album covers. I tried to start a conversation but was nervous and could not find anything to say. What is it about fame? Just being in the same room as the famous makes us feel good or even glamorous by association. It never ceases to amuse and entertain me when I go somewhere and people want to meet me because they think I am famous. I may be somewhat famous to some *AJNR* readers, but that is the extent of my 15 minutes (or perhaps, just my 15 seconds). Are famous individuals that different from you and me?

Part 2: Some Facts

What makes a person famous? Are the brains of famous people different from those of the rest of us? It is well known that Einstein's brain was removed 7 hours after his death and preserved in formalin.[1] Its examination showed differences in the parietal operculum, inferior frontal gyrus, and frontal lobe, leading to speculation that because of this, his neuronal connections were faster and more efficient. Histology indicated that some parts of his brain had more glial cells than those found in controls, but what this really means is uncertain. It is a fact that many wanted a part of his famous brain, a story entertainingly told in Michael Pattertini's book *Driving Mr. Albert: A Trip Across America with Einstein's Brain.*[2] Lenin's brain was also removed to gain insight into the genius of communism (American spies tried to get it but were unsuccessful), and it showed a greater-than-average number of pyramidal neurons, which the famous German neuropathologist Oskar Vogt thought accounted for Lenin's greater intellect.[3]

After Vogt's report, the Russians funded the Moscow Brain Research Institute, which houses a collection of their most famous brains (including those of Pavlov and Mendeleev), to study the neuroanatomic basis of exceptional mental capacity and talent.[4]

Data were kept secret until the 1990 and, later when revealed, showed that famous brains had nothing different from yours or mine.

In Japan, at the University of Tokyo, more than 120 brains of famous Japanese prime ministers, novelists, and scholars are kept in a sort of museum and studied to gain insights into what made these individuals who they were.[5] The Japanese are very respectful of the dead and prefer to keep brains intact rather than cutting them. They have found that these famous brains weigh more than those of common individuals. In the United States, in San Diego at the Department of Radiology of the University of California, one finds the Brain Observatory.[6] There, more than 1000 donated brains are being studied, the most famous belonging to patient H.M., an individual incapable of forming new memories since neurosurgery for seizures nearly 50 years previously damaged both hippocampi. Other famous institutes housing brains include the Wilder Brain Collection at Cornell University (Dr Wilder was a Civil War surgeon and animal biologist; his collection contains brains of some infamous individuals), the C.U. Ariens Kappers Brain Collection at the Netherlands Institute for Brain Research (it contains mostly animal but also some human brains), and several collections at the Vogt Institut fur Hirnforschung at the University of Dusseldorf in Germany.[7] When it comes to donating a brain to science, famous individuals as a whole tend to do it more willingly than others. This is probably an extension of their ego and the desire to remain the center of attention even after death. German dramatist, Johann Von Schiller said, "Of all the possessions of this life fame is the noblest; when the body has sunk into the dust the great name still lives."[8]

Because, after all, famous brains are not significantly different from those of common individuals, it is interesting to study how living brains react to fame. It has been postulated that entire brain networks are needed to effectively recognize objects and faces (easier when they are famous). Some studies demonstrate that more than 1 dedicated network is needed for recognition—that is, to recognize a face, visual and vocal networks need to be activated and work together (called dynamic cross-modal networks).[9] Face recognition depends on intact networks located in the temporal and occipital lobes and anchored by the hippocampi, all working on a parallel basis. Conversely, in 1 study, famous faces elicited a response of only a few neurons located in the medial temporal .[10] The faces of President Clinton, the Beatles, the Simpsons, Jennifer Aniston, and

Michael Jordan, among others, were shown to 8 subjects in 21 different sessions. Although all neurons fired when Aniston was shown alone, pictures of her with Brad Pitt resulted in suddenly quiet neurons, an observation difficult to explain. Halle Berry dressed as Catwoman also elicited consistent neuronal firing independent of the subject's sex. Cells that respond to famous human faces have little baseline activity but are turned on rapidly when a familiar face is seen. Thus, some studies suggest that entire neural networks may not be needed for recognition of familiar faces.

It is also amazing how face recognition can erase discrimination.

In 1 study, a set of white and black subjects were shown pictures of white and black unknown individuals. With repeated viewing, the brains of whites showed less activity when seeing other whites; the same happened to black subjects but not vice versa.[11] In a different study, white females were shown faces of white and black individuals (neutral expression and all obtained from old college yearbooks) and were asked to associate these with words.[12] Black faces elicited a "pro-black bias effect" and were associated with words such as "cancer" and "bomb." The same investigators also performed a similar experiment, but this time showed black faces of famous (not infamous) individuals to the subjects, and amazingly, all negative biases disappeared completely, implying that familiarity (and fame) erased the initial feelings.

[13] Not surprisingly, both articles elicited considerable concern about stereotyping and prejudice and were subject to intense press coverage.

It is also interesting that the more you see a face, the less you use your brain to recognize it. Thus, little effort is needed to recognize the famous (probably explaining why only a few neurons are used). This is particularly true if faces belong to one's own race. The so-called "other race effect" states that as long as the face you are seeing is of your same race, your brain will learn very rapidly to recognize it and will activate less each time you see it, something that takes longer to occur when shown faces of other races.[14] The opposite is called "cross-race effect"—that is, more effort is needed to recognize faces and expressions of individuals of races other than your own.[15] The implications of this concept are interesting as we become a more multicultural society. Interestingly, alcohol decreases our ability to recognize familiar faces, and all individuals start looking the same. Therefore, if you get assaulted while intoxicated, even if you recognize the perpetrator in a line-up, your testimony may not hold up in court.

Part 3: More Anecdotes (Other Famous Faces I Have Been Able To Recognize)

Jack White (of the White Stripes). During a recent trip, I sat in a waiting room at the Detroit airport. Beside me was this guy dressed in black, and upon recognizing him, I said, "Jaackkk Whiiittee!" He looked at me, gave me the smallest of smiles, got up, and walked away. He is taller than I previously thought.

Richard Nixon. As a child, my family and I arrived very late at the Key Biscayne Hotel and Villas for a vacation. The dining room was empty except for the President and his wife, who were eating. My father wanted me to go say hello, but I was too shy to do so. The next year, we arrived late again only to find that there was no space for us in the hotel. We ended up spending the night in Mr. Nixon's villa; it was nothing special.

The Boss. With my family, I was having brunch at a restaurant of the Nasher Museum of Duke University during Parents' Weekend when Mr. Springsteen walked in (his daughter is a student there).

He waited in line nearly 30 minutes for a table, just like us mortals.

Mr. Springsteen is taller than I thought and looks exactly like he does in pictures. We could not stop staring at him, but because he wore dark sunglasses throughout the meal, we have no idea if he actually saw us doing this.

Part 4: Famous Last Words

I hope I haven't bored you.

—Elvis Presley, closing remarks during his last press conference

REFERENCES

1. Albert Einstein's Brain. http://en.wikipedia.org/wiki/Albert_Einstein%27s_brain. Accessed February 16, 2012
2. Pattertini M. *Driving Mr. Albert: A Trip Across America with Einstein's Brain*. New York: Dial Press; 2001
3. Oskar Vogt. http://en.wikipedia.org/wiki/Oskar_Vogt. Accessed February 16, 2012
4. Vein AA, Maat-Schieman ML. **Famous Russian brains: historical attempts to understand intelligence.** *Brain* 2008;131:583–90
5. Kurtenbach E. Famous brains preserved for posterity: research: Japanese scientists hope to gain insight into the mind of great thinkers. *Los Angeles Times*. January 20, 1991. http://articles.latimes.com/1991–01-20/news/mn-691_1_famous-brains. Accessed February 16, 2012
6. THE BRAIN OBSERVATORY. http://thebrainobservatory.ucsd.edu. Accessed February 16, 2012
7. Other brain collections. http://brainmuseum.org/explore/other brain-coll.html. Accessed February 16, 2012
8. All the Best Quotes. http://chatna.com/theme/fame.htm. Accessed February 16, 2012

9. Joassin F, Psenti M, Maurage P, et al. **Cross-modal interactions between human faces and voices involved in person recognition.** *Cortex* 2011;47:367–76
10. Quiroga RQ, Reddy L, Kreiman G, et al. **Invariant visual representation by single neurons in the human brain.** *Nature* 2005;435:1102–07
11. **Human kinds in the brain: an MRI scan of racial perception.** http://www.davidberreby.com/human_kinds_in_the_brain__an_mri_ scan_of_racial_perception_41324.htm. Accessed February 16, 2012
12. Phelps EA, Cannistraci CJ, Cunningham WA. **Intact performance on an indirect measure of race bias following amygdala damage.** *Neuropsychologia* 2003;41:203–08
13. Phelps EA, O'Connor KJ, Cunningham WA, et al. **Performance on indirect measures of race evaluation predicts amygdala activation.** *J Cog Neuroscience* 2000;12:729–38
14. Vizioli L, Rousselet A, Caldara R. **Neural repetition suppression to identity is abolished by other-race faces.** *Proc Natl Acad Sci U S A* 2010;107:20081–86
15. Elfenbein, HA, Ambady N. **When familiarity breeds accuracy: cultural exposure and facial emotion recognition.** *J Pers Soc Psychol* 2003;85:276–90

Decisions:

OVERWHELMED BY CHOICES

> *It is our choices that show what we truly are, far more than our abilities.*
>
> J.K. Rowling, *Harry Potter and the Chamber of Secrets*

What should I choose? Pizza or Chinese? The blue or the white one? We wish all choices in life were this simple.

"Choice" is defined as the ability to exercise control, and whether we are able to truly choose is greatly affected by the initial perception of being able to do so. If one believes in having choices, there is a personal benefit even in the absence of exercising them. In a Capitalist economic system like ours that emphasizes individualism, choosing is an extremely important aspect of everyday life.

Our perception of choices is directly influenced by the myriad products available (which in reality are made by a dwindling number of corporations in the same place by using the same raw materials, resulting in actually fewer choices). The concept of "choice" as we now know it started in the 17th and 18th centuries in Western Europe. Ironically, Collectivism (which offers fewer choices) also started in Europe (by Karl Marx) as a reaction to the Industrial Revolution and the idea of Individualism.

In the West, our desire to choose separates us at an early stage from our parents, whereas in other cultures, particularly Asian, younger family members always consider their parents' opinions when making a choice. It is now said that between 18 and 25 years of age, we pass through a period of self-discovery and choice. This period of life is becoming more important and obvious as our children stay at home later until they finally choose what to do with their lives. Today, the countries where choice is considered a most important human right are the United States, Australia, and Great Britain. In the United States, choice is a matter of principle and practice, while in the rest of the world, Collectivism diminishes its importance. The only country where perhaps Collectivism and Capitalism coexist is Japan, and even there Collectivism takes precedent like it does in most Asian countries.

In the United States, we take "freedom of choice" as an undeniable right. This even extends to spirituality, and in 2009, a Pew Institute survey showed that 50% of Americans chose to change religions at least once during the lives![1] Similarly, the right to choose whom we marry is probably overemphasized and finally not that important. When American couples were asked 20 years later how they felt about each other, nearly all stated that "love" was no longer present in the relationship. However, when Indians, for whom a spouse was chosen and unknown before the wedding, were asked the same question, 60% stated they loved each other.[1] Of course, this can also reflect what we understand as "love." Although choices such as religion and marriage are influenced by endless variables, neuroimaging provides a glimpse into how the brain makes simpler choices.

Brain activity before making a choice begins in either the prefrontal or parietal cortex.[2] These parts of the brain achieve their final size between 20 and 30 years of age, and that is perhaps why most young adults really cannot choose "wisely" until about 25 years of age (vide supra). Most choices involve motor elements.

The prefrontal cortex basically makes a cost/benefit analysis before telling the basal ganglia to plan a movement. One of the most common daily choices we face is which hand to use. In this situation, the posterior parietal cortex initiates competitive but parallel impulses for both hands, and once one process reaches a threshold, the corresponding hand does the reaching. The more ambiguous the target is, the greater the preparation time the parietal lobes need. A study using fMRI showed that there may be up to a 7-second delay between cortical activation and the final push of a button.[3] These studies bring up 2 important aspects: The brain always has a plan B, and often there is time to avert an action. Because choosing is directly related to evolutionary adaptation and keeps us being the supreme animal on earth, it is worth taking some time to explore other aspects related to how we choose. The subspecialty that investigates how we choose is called "decision neuroscience," and as expected, economists and business managers are very interested in it.

Before making a decision, the brain must learn to predict when and where rewards based on choice will occur.[4] This ability is basically acquired by observation. The accumulated information is processed in several parts of the brain (mainly the orbitofrontal cortex and amygdalae, which makes sense because these are very primitive parts of the brain) that assign value and relevance to our choices. It is thought that wanting something is mediated by dopamine, while liking something depends on opioid neurotransmission that occurs mostly at the nucleus accumbens. Finally, reversal-learning, planning, and avoiding impulsive choices lead an individual to select the best choice. Compulsive gamblers seem to have a problem at this last level. Some drugs and mood disorders also affect choosing at this last stage. Each choice is assigned a different value, and on the basis of these values, we humans choose. If one learns that regret accompanies a choice, this specific choice ceases to be selected.[5] If one chooses the same option over and over again, satiation takes place and we lose our interest in that particular choice.

Mental priming is a prerequisite to choosing. A "prime" is an item of information or implicit memory stored in our brains that is enhanced because of a particular stimulus. The industry uses advertisements to create these primes that later predispose customers to make certain choices when shopping. Even words are not free of priming. Free-association word activities are not entirely "free" because one word always primes a link between it and the subsequent one (the word "dog" usually primes the word "cat" and vice versa). This is called "semantic priming," but priming can also be perceptual and conceptual.[6] An example of perceptual priming is being asked to complete a word. If you are shown "airpl . . . ," you can easily guess the rest of it ("airplane").

Conceptual priming is seen when 2 concepts belong to the same category, such as "table" and "chair." Priming can be positive (it increases the speed with which we choose) or negative (slows it).

Positive priming occurs by a spread of brain activation, but negative priming has not been explained. Repetition reinforces positive priming and leads to faster choosing. As we become more primed to respond to something, less neural processing is involved.

When neuroimages are interpreted, clinical information and previous knowledge of radiology help us first to choose the areas of the brain likely to be involved by a particular disease (called "top-down instruction"). Non physicians do not choose these same specific areas of an imaging study but look everywhere (called "bottom-up salience"). Using top-down methods, experienced mammographers can detect cancers within 1.0 second of being shown a study.[7] Repetitive priming results in a "look-detect-scan" pattern rather than the "scan-look-detect" pattern. In radiology, we talk about the importance of "Gestalt" recognition and interpretation of findings (equivalent to "look-detect-scan").

In one study, neurologists who were shown a series of brain CT scans with infarctions often stared at an anterior cerebral artery territory infarct for an average of 11.5 seconds before choosing their diagnosis.[8] This study done by neurologists claims that this delay occurs because head CT studies are much more complex than, say, mammograms (I do not agree with this claim). We interpret imaging studies by narrowing down our choices to offer a credible differential diagnosis.

Similarly when faced with an immense number of products on a supermarket shelf, we narrow down our choices and disregard all other products. Economists know this well: You can only put so much stuff on a shelf before humans begin to ignore most of it (this is called the "more is less" principle). For unknown reasons, humans choose in sets of 3. At the grocery store, you will probably narrow down your choices to 3 and end up buying 1 of these. I do the same when presenting unknown cases to our trainees and ask them to give me the 3 most important differential diagnoses and finally to choose 1. Because we never provide thousands of differential diagnoses, choosing may not be as difficult as going to your local supermarket, which carries an average of 39,000 items.[9]

What makes the most difference between these 2 choosing situations is expertise. I have the expertise to offer you the 3 most likely diagnoses but not the expertise to choose among dozens of dishwashing detergents. The expert mind works at a more granular level and develops the ability to exclude unwanted choices rapidly. Because we are not experts in everything, we must heed the advice of other experts. For example, how do you know that a certain article in the *American Journal of Neuroradiology* is the best choice for you? We now provide a rating system with which expert readers can rate articles so that when you search for articles on "penumbra imaging," others will probably have already rated these, letting you know which are best, and thus you will not be overwhelmed by choices. Recommendations and categorizations make choosing easier and wiser, and restrictions lead to choosing correctly within a framework. Famous jazz player, composer, and overall musical ambassador Wynton Marsalis said: "You need to have some restrictions in jazz. Anyone can improvise with no restrictions, but that's not jazz. Jazz always has some restrictions.

Otherwise it may sound as noise."[1] I would like to extend this concept to our daily image interpretations: Without restrictions, categorizations, recommendations, and a narrow focus, we may overwhelm our colleagues and our choices may be interpreted as just noise.

REFERENCES

1. Iyengar S. *The Art of Choosing.* New York: Twelve; 2010
2. Oliveira FL, Diedrichsen J, Verstynen T, et al. **Transcranial magnetic stimulation of posterior parietal cortex affects decision of hand choice.** *Proc Natl Acad Sci U S A* 2010;107:17751–56
3. Exploring the Mind. Brain scans can reveal your decisions 7 seconds before you "decide." http://exploringthemind.com/the-mind/brainscans-can-reveal-your-decisions-7-seconds-before-you-decide. Accessed June 22, 2012
4. Diekhof EK, Ralkia P, Gruber O. **Functional neuroimaging of reward processing and decision-making: a review of aberrant motivational**

and affective processing in addiction and mood disorders. *Brain Research Reviews* 2008;59:164–84

5. Coricelli G, Critchley HG, Joffily M, et al. **Regret and its avoidance: a neuroimaging study of choice behavior.** *Nature Neuroscience* 2005;8:1255–62
6. Priming (psychology). Wikipedia. http://en.wikipedia.org/wiki/ Priming_(psychology). Accessed June 22, 2012
7. Kundel HL, Nodine CF, Krupinski EA, et al. **Using gaze-tracking data and mixture distribution analysis to support a holistic model for the detection of cancers on mammograms.** *Acad Radiol* 2008;15:881–86
8. Matsumoto H, Terao Y, Yugeta A, et al. **Where do neurologists look when viewing brain CT images? An eye-tracking study involving stroke cases.** *PLoS One* 2011;6:e28928
9. The Voice of Food Retail. Supermarket facts. http://www.fmi.org/ facts_figs/?fuseaction_superfact. Accessed June 22, 2012

CROSSING THE RUBICON

If you commit a crime, it is well-known that it is better to face the judge in the morning than later in the day. Studies show that criminals are less likely to be convicted if their case is heard in the early morning than in the late afternoon.[1] Although the above scenario will hopefully never affect any of us neuroradiologists, I do wonder how it affects our daily work. Am I more likely to make the wrong decision at the end of a long day in the reading room?

As an editor, am I more likely to accept or reject articles late in the evening?*

Repeated judgments and decisions exhaust one's mental resources, especially those related to executive functions and thus influence the way we make choices leading to intuitive rather than rational decision-making. Lesson: never buy a car late in the afternoon; do it early, during the weekend, and after sleeping well the night before. Individuals who buy their cars when tired are more likely to accept whatever the manufacturers say about their products.[2] Car salespersons know how to play the game as they go back and forth to their managers, each time forcing you to make new decisions (paint type, type of wheels, leather or standard cloth, and so forth) that wear you down so that when the time to make the final decision to buy the car comes, you are exhausted and just buy it as is. Candy manufacturers are also experts at decision fatigue, and they generally place their (junkier and cheaper) products by the supermarket cashier, so when one is exhausted from shopping, poor decision-making kicks in and one ends up buying all sorts of junk food. Making more decisions during one's day does not make one better at them; on the contrary, it makes for worse choices. The impulsive decision to answer a controversial e-mail when fatigued is nearly always the wrong one; something that I learned the painful way.

The Rubicon decision model states that there are 3 stages to decision-making: first a predecisional (orientation) stage and last a postdecisional stage (also called emergence and reinforcement stages), which are separated by a Rubicon (named after the river in northern Italy that Caesar crossed resulting in a civil war between Gaul and Rome). Thus, the expression "crossing the Rubicon" refers to making an irreversible decision. When one makes the decision to click on the approve button for an imaging report, the "imaging interpretation Rubicon" has been crossed. Crossing the Rubicon is one of the most exhausting stages of decision making, so it should not be surprising that we radiologists feel drained by the end of the day.

The Rubicon process is even more complicated when a clear cut decision cannot be made and one must make trade-offs (called the conflict stage). Compromise and trade-offs in the predecisional stage are also exhausting. If one cannot afford to buy an item, trade-offs lead to exhaustion and not uncommonly one ends up making a decision based only on price rather than other important features such as quality. Marketing strategies also take advantage of this type of fatigue to time the sale of items when you are indecisive. Fatigue can also work in other ways when you are trying to decide. Many choices, conversely, may not make decisions easier but rather lead to no decisions at all.

Some studies have shown that medical-decision errors often occur under the effects of stress and fatigue. In addition, medical teams are unlikely to discuss these errors and to accept the role that decision fatigue played in them. However, airline cockpit crews are prone to report these errors to help improve their performance.

[3] In the 1950s, commercial airline accidents dropped significantly when it became obvious that many errors arose from incorrect decision-making by pilots and crew. Critical personality features that airlines look for in their pilots include the ability to learn from errors, recognize them in others, and contribute to solving them. Today, more than 7 generations of airline crews have been trained with these features in mind, and aviation accidents are at an all-time low.

Work hours may be limited to avoid fatigue and errors in decision-making. We are all familiar with the application of this concept in regard to the maximum hours that trainees in medicine can work (80 per week). These hours include hospital day work, call hours, and time spent in academic activities related to training. Additionally, any other activity for which "attendance is strongly encouraged" (such as off-hours journal clubs) needs to be counted within the allowed 80 hours. When a resident travels to present work at a meeting, this travel also counts as active duty time. Why does this requirement apply to trainees and not us, full-fledged radiologists? I spend about 10–12 hours per day at the hospital, plus an additional 2–3 per day working on *AJNR*, and countless others doing service and volunteer work. Yet, the ultimate responsibility of making a decision in reporting imaging studies falls on me. Are younger individuals more resilient to fatigue than us older ones? Airline rules apply equally to senior and less experienced crew members.

The origin of limiting the hours trainees can work arose from recommendations made by the Institute of Medicine in 2008 with the intention of limiting trainee fatigue and decision-making errors.

Younger trainees are thought to be more affected by fatigue; therefore, they are allowed to work fewer hours than senior ones (16 versus 28 continuous hours).[4] A study published in the *New England Journal of Medicine* points out how these new rules made a bad situation even worse. Data from a survey completed by more than 6200 residents showed that residents believed that their schedules, education, and quality of life were actually worse and that patient care was suffering after implementation of the new hour policy in 2011.[5] This finding makes perfect sense because while it relieved fatigue in first-year trainees, the bulk of the work was shifted to more senior residents who actually compose the largest pool. Another study reported that despite the new work hour requirements, the amount of sleep residents got did not improve.[6] If limiting duty hours was an attempt to improve patient safety, this also failed and intensive care unit stays are more likely than before the reform.[7] To be fair, I must mention that many other studies (too many to quote here) have shown that errors in laparoscopy and electrocardiogram interpretation, unnecessary prolongation of procedures, and lesser quality physical examinations occur more often when chronically sleep-deprived residents have been involved.

Why limiting duty hours for our trainees failed is difficult to understand. Airline pilots are allowed only 8 hours of flight time in a 24-hour period, and all pilots must rest a minimum of 8 hours between assignments.[8] For flights lasting more than 12 hours, adequate sleeping facilities outside the flight deck must be provided.

Airlines also provide their pilots with "fatigue" training. If pilots nap, why not our residents? The Accreditation Council for Graduate Medical Education actually says, "Strategic napping is strongly suggested in the program requirements . . . and should not be scheduled but based upon resident fatigue."[9]

Although intuition tells us that working overtime leads to fatigue, this is not always the case. In a survey of 4000 other workers, the amount of non-extreme overtime did not correlate with fatigue.

[10] Moreover, overtime workers had favorable characteristics such as high decision latitude, high job variety, high job demands, and, most important, a high job satisfaction. It appears that personal motivation is what leads these workers to work more while maintaining high satisfaction. This, however, may not be the case in Asia where doctors are exposed to extreme overtime (more than 60 additional hours per week) and chronic fatigue is common.[10]

The key to the failure of the new resident work hour requirements to improve fatigue and decision-making errors may lie in our working environments. In an above-cited study, a positive psychosocial environment was critical to ameliorate the effects of fatigue.[10] This may be why all of these policies work better in the cockpit than in our reading rooms. Cockpits are highly structured, relatively quiet, free of inconsequential interruptions, and respectful places to work, while our reading rooms with their myriad interruptions by telephone, secretaries, clerks, technologists, students, and other physicians have become stressful, nightmarish environments in which it is difficult to complete the assigned tasks and concentrate on making the correct decisions.

Below are some of the most common problems leading to decision-making mistakes and how I think they relate to radiologists:

1. Not taking enough time. A common problem in my reading room in which 1 fellow, 1 resident, and 1 attending interpret upwards of 100 head CTs and MRIs during the workday (spine and ear, nose, and throat studies are interpreted in different reading rooms that have the same problems).
2. Lacking a peaceful environment (see above). Business folks put it in different language: wallowing in chaos.
3. Not considering priorities. In marketing, this is akin to not doing what is best for you, neglecting your values, avoiding the truth, procrastinating, and ignoring what is right.
4. Learning how to say no. I think that learning when to stop interpreting studies and realizing and accepting fatigue are essential to avoid mistakes. Item 3 comes into play here. In the late afternoon, one must prioritize the studies that need to be read and leave others for later. I have learned something else:

There is no reason why a case that one does not understand must be immediately interpreted. Now, I place a preliminary report with a note stating that I would like to think about it for a while. This pause allows me to go to my office and do some research, which has always been a good decision, resulting in more intelligent reports. This is what experts call "timing" your decision.

Well, most of this advice is easier said (written) than done.

Fatigue will always affect the way we make decisions, and as our existing medical delivery systems come under stress, more work and fatigue will be the rule rather than the exception. Remember that a good and timely decision does not guarantee a good outcome.

Decisions must be plastic; they should be able to change and mold when new information becomes available. An irrevocable decision to cross the Rubicon always carries with it the risk of the Ides of March.

REFERENCES

1. Danziger S, Levav J, Avnaim-Pesso L. **Extraneous factors in judicial decisions.** *Proc Natl Acad Sci U S A* 2011;108:6889–92
2. Levav J, Heitmann H, Herrmann A, et al. **Order in product customization decisions: evidence from field experiments.** *J Polit Econ* 2010;118:274–99
3. Sexton JB, Thomas EJ, Halmriech RL. **Error, stress, and teamwork in medicine and aviation: cross sectional surveys.** *BMJ* 2000;320:745–49
4. Fodeman JD. **The new resident duty hours fail.** *National Review Online*. August 12, 2012. http://www.nationalreview.com/critical-condition/ 313022/new-resident-duty-hours-fail-jason-d-fodeman. Accessed March 19, 2014
5. Drolet BC, Christopher DA, Fischer SA. **Resident's response to dutyhour regulations: a follow-up national survey.** *N Engl J Med* 2010; 363:e34
6. Schumacher DJ, Frintner MP, Jain A, et al. **The 2011 ACGME standards: impact report by graduating residents on the working and learning environment.** *Acad Pediatr* 2014;14:149–54
7. Block L, Jarlenski M, Wu AW, et al. **Inpatient safety outcomes following the 2011 residency work-hour reform.** *J Hosp Med* 2014; 9:347–52

8. Federal Aviation Administration. Fact sheet—pilot flight times, rest, and fatigue. January 27, 2010. http://www.faa.gov/news/fact_sheets/ news_ story.cfm?newsId_6762. Accessed March 19, 2014
9. Accreditation Council for Graduate Medical Education. Frequently asked questions: ACGME common duty hour requirements. https://www. acgme.org/acgmeweb/Portals/0/PDFs/dh-faqs2011.pdf. Updated June 18, 2014. Accessed September 19, 2014
10. Beckers D, van der Linden D, Smulders P, et al. **Working overtime hours: relations with fatigue, work motivation, and the quality of work.** *J Occup Environ Med* 2004;46:1282–89

Techie stuff:

ICONSENT

In medicine, a "consent form" is the legal instrument through which we must give our patients sufficient information (positive, such as benefits, and negative, such as risks and complications) regarding any treatment or procedure they will receive. The idea behind it is that an informed patient can accept or more importantly decline a treatment (for personal and/or religious reasons) even if the physician disagrees with this decision. The key word here is "physician." A consent form, under most circumstances, must be administered only by a full-fledged physician, never a medical student, nurse, or technologist. The attending physician can (and does in many academic centers) delegate obtaining consent to a resident or fellow (who already is an MD). In the United States, minors cannot give consent and their parents or legal guardians must give it.* Patients in extreme emergency situations and those with limited cognition are exceptions; and when no family member or legal guardian is available, 1 or more physicians may sign the patient's consent form. As we get older and move into specialized care institutions far away from our families, the caregiver may consent to emergency treatments. The Caregiver Consent Form must be prepared in advance; a lawyer is not needed in the decision making process. For parentless children, a similar form can be used. Consent forms from parents, grandparents, and others are available in most large institutions.

The most common consent form used is, however, the generic one. Regardless of their specifics, all consent forms must meet certain minimum legal standards. Any impairment of reasoning faculties and/or judgment (including previous sedation) makes it impossible (and illegal) to administer the consent form, regardless of its type. Waivers of consent may also be obtained and are not uncommonly used by large institutions such as the Army when a treatment involves minimal risk, benefits the patient, advances medicine, and is carried out under laws established by the US Food and Drug Administration.[1] The need for consent is so ever-present that there are commercial companies that specialize in designing and administering these forms.

Access to the Internet and medical knowledge has considerably changed many aspects of consent. Until a few decades ago, medical treatment was administered following the concept that "doctors know better." This idea originated in Greece and follows the precepts of the Hippocratic Oath.[2] Many of us become irritated when patients try to steer their treatments (coil embolization versus clipping of intracranial aneurysms is a typical example) on the basis of information found on the Internet because we have been brought up to believe in the Hippocratic Oath (ie, we know better). This concept did not really change until the 18th century, when doctors began to believe that sharing as much information as possible with patients was beneficial, but in the end, physicians always made the most important and final decisions.

The idea the "doctor knows better" has been called "benevolent deception," and it has been fought against since the mid-1800s. In the United States, the most important aspect of consent is "what is being said" rather than "who is saying it" and "where it is being said" (this may not be the case in other cultures and countries).

As we now know it, the consent form is a recent invention and stems from the consequences of various unethical (to say the least) situations during and around World War II. After the war trials against illegal human experiments by Nazi physicians, the Council for War Crimes published the "Nuremberg Code."[3] This set of rules defines legitimate medical research and is accepted by the Declaration of Helsinki and the US Department of Health and Human Services and is incorporated into the law in many states and countries. One of the most important aspects of the "Nuremberg Code" is informed consent without coercion. Violations of the Code continued after the War even in the United States. Perhaps the best known is the "Tuskegee Syphilis Experiment."[4] This experiment (if one can call it that) extended for 40 years (up to 1972) and was "administered" by the US Public Health Service in Tuskegee, Alabama. In it, the natural progression of syphilis was assessed while infected patients thought they were getting the appropriate medical treatment. Six hundred poor African American agricultural workers were recruited, and 400 who had syphilis went untreated (they were given free burial insurance by the government).

Remember that 15 years after the beginning of the "experiment," there was irrefutable scientific evidence that penicillin (widely available by then) was the standard treatment for syphilis.

Although this is not the only occasion of human rights violations, it is certainly the most infamous one, and in 1978, it led to the "Belmont Report," which sets the guidelines for the protection of subjects in clinical and research trials in health care.[5] The report led to the creation of the Office for Human Research Protections and Institutional Review Boards (IRBs) in medical schools, academic centers, and hospitals.

IRBs are decentralized committees that review and monitor biomedical research in humans. IRBs themselves are overseen by the Office for Human Research Protections. Before becoming an IRB member at any institution, any conflict of interest (such as working for the industry as a consultant) must be disclosed. IRBs must comprise at least 5 experienced individuals (both male and female), have representatives of different professions (scientists versus nonscientists), and include community members.[6] All research projects and, in many institutions, all publications must be granted permission by an IRB. These IRBs approve research projects only when bona fide consent will be obtained from all participants.

When the project is closed, most IRBs require notification and summary of the results.

Most of us who have been (or are) involved in research know how difficult and lengthy the process of IRB approval is. Many blame the relative decline of US research on this while other countries with less complex approval processes are making headway in research. To ease the process, many IRBs offer exemptions. In medicine, the most common exemption is for research that involves the analysis of already-existing data as long as the identities of the subjects are protected. For this, some IRBs have special shorter forms while others demand that their long forms be completed.

For most exemptions, consent from individual subjects is not required. The problem with IRBs is that data obtained from patients are so closely guarded that access is not available to other researchers who would benefit from them. Just try getting your own data after participating in a research project that has been completed. Even worse, try getting your own medical record released.

The owner of the medical record is not the patient but the health service provider who created it, and similarly, the owner of data collected during research is the institution or company funding the project and not the subjects.

Research data are kept in "information silos" understandably guarded from prying eyes but also fragmented. Similar to grain silos that house one type of product, data collected are mined only for proving or disproving a specific hypothesis, and all other information contained in the silo is not used. In this era of fast computing, data transfer, and crowdsourcing and sharing, this process may not be the best way to advance science. "Open source" medicine and research are coming our way, and we need to adapt more than our consent forms to take advantage of them.

Apple (Cupertino, California) and Google (Mountain View, California) already keep track of an enormous amount of personal data; Microsoft (Bothell, Washington) keeps track of all data transmitted by using their products such as Outlook for e-mail and calendars. Very soon, science will not survive without data sharing, integration, and networking. It could be that the consent form that was created to protect us is now, in its current form, detrimental to science.

John Wilbanks has created the WeConsent.us Web site and data base (http://weconsent.us/).[7] Mr. Wilbanks said, "All too many observations lie isolated and forgotten on personal hard drives and CDs, trapped by technical, legal and cultural barriers."

A critical and innovative aspect of this idea is the use of a special consent form that states that if kept anonymously, your (and my) medical data (particularly health and genomics) can be used by third parties as long as our identity remains protected. This consent is called a "Portable Legal Consent" because you carry it with you, and you attach it (thus its portability) to any data you want to donate. Think about it as having a consent form in your iPhone (Apple) and electronically transmitting it when you need. Personally, I would not mind sharing my medical data if my identity is protected, but I cannot do this because I do not own it! Mr. Wilbanks stated that we need to move from information silos to "information commons."

Vanderbilt University (in collaboration with Northwestern University) has started a DNA biorepository and combining it with electronic medical records, an information commons expected to shed light on diabetes, Alzheimer disease, and heart disease is being built.[8] When asked, nearly 95% of patients state that they would be willing to share their medical data.[9] Applications for the iPhone (i.e., MyMedical) that allow you to keep your own medical record and share all or parts of it are available. The Eatery application allows you to photograph what you eat and share it (anonymously) with other users to try to improve your eating habits. The goal of the WeConsent.us Web site is to get 100,000 individuals in its first year (and 1 million in 5 years) to donate their medical data, which will then be available for analysis by mathematicians and other scientists.

The future of medical research lies in its power, and its power lies in numbers. However, this power can only be realized if we own our data and we consent to share it. Data accumulated with time do not have to wait to be uploaded and shared but should be dynamically shared in real time. Imagine carrying your own consent form in your mobile device and attaching it to newly available data that you can share when you want to. This consent form could be malleable and would adapt to different needs and situations, taking advantage of the incredible interaction possible on the Web. I think that the time for the iConsent is here.

* In other countries (especially England), the Gillick standard states that a child younger than 16 years of age may, under certain circumstances, be judged mature enough to consent.[10]

REFERENCES

1. McManus J, Mehta SG, McClinton AR, et al. **Informed consent and ethical issues in military medical research.** *Acad Emerg Med* 2005;12: 1120–26
2. Informed consent. Wikipedia. http://en.wikipedia.org/wiki/Informed_consent. Accessed October 24, 2012
3. Nuremberg Code. National Institutes of Health. http://history.nih. gov/research/downloads/nuremberg.pdf. Accessed June 20, 2012
4. Tuskegee syphilis experiment. Wikipedia. http://en.wikipedia.org/ wiki/Tuskegee_syphilis_experiment. Accessed October 24, 2012
5. Sims JM. **A brief review of the Belmont Report.** *Dimens Crit Care Nurs* 2010;29:173–74
6. Institutional Review Board. Wikipedia. http://en.wikipedia.org/wiki/Institutional_Review_Board. Accessed October 24, 2012
7. John Wilbanks: let's pool our medical data. http://www.ted.com/ talks/john_wilbanks_let_s_pool_our_medical_data.html. Accessed October 24, 2012
8. The eMerge Network. http://emerge.mc.vanderbilt.edu. Accessed October 24, 2012
9. Most consumers willing to share personal data. NorthPoint Domain. http://www.northpointdomain.com/default/index.cfm/expertiseand-insight/iq360-insight/industry-spotlight/most-consumerswilling- to-share-personal-data. Accessed October 24, 2012
10. Gillick competence. http://en.wikipedia.org/wiki/Gillick_competence. Accessed October 24, 2012

FROM HARD DRIVES TO FLASH DRIVES TO DNA DRIVES

The word is now a virus.
William S. Burroughs, *The Ticket that Exploded*

Recently there has been another round of controversial news regarding genetically modified organisms (GMO). Perhaps the best known debate on these centers on corn. A recent French study showed severe kidney and liver abnormalities in rats that were fed this corn for up to 2 years.[1] Immediately afterward, Russia banned the use of this seed and the corn it produces. Because other studies have not confirmed this finding, the American media immediately released news stories stating that the French study was flawed and unscientific and that it represented just another round of propaganda by individuals who oppose GMO and the companies that produce the seeds (which are mostly American).

[2] Salmon, with growth hormones that have been altered so that they not only grow faster but never stop growing, has also been in the news. Salmon is the third most-eaten seafood in the United States according to the National Fisheries Institute, and most of it is flown in from Chile, so growing enough of it here to feed Americans may actually be a good thing for the environment, even if its genes have been modified.[3] All of these situations involve inserting or altering a specific gene in plant or animal deoxyribonucleic acid (DNA); thus, the genetic material in those organisms still serves its original purpose. However, what happens if we take our DNA, reconfigure it, and use it for something completely different from that for which it is intended? Cutting edge genetic engineers are now synthetizing DNA so that it contains information much like a computer hard drive or solid memory chips. The capacity of DNA as a storage medium is staggering:

All of the information contained on the entire Internet would fit into a device smaller than 1 cubic inch!

As our need for high-capacity information storage continues to increase, several researchers have begun to explore the possibility of using DNA for this purpose.[4] The very fabric of life uses a binary code, but instead of the 1s and 0s computers use, the code in our DNA is composed of 4 letters: A, G, C, and T (adenine, guanine, cytosine, and thymine), which are paired into 2 nucleotide bases: A-T and G-C (hence a form of binary code). By changing the order of these 2 base nucleotide pairs, one can encode all different types of information in the same way a computer does by changing the order of 1s and 0s. Each nucleotide may encode 2 bits of information, and 1 g of single-stranded DNA can store 455 exabytes. One exabyte is equal to 1000 petabytes; 1 petabyte is equal to 1 quadrillion bytes, and so on. What this means is that in 1 g of single-stranded DNA, one can potentially store the equivalent to 250 million DVDs! Computer chips are "planar" storage devices (obvious from their shape). One way to improve the capacity of a computer chip is to put several layers of circuits in it (making it 2D), but because DNA is 3D, it offers much more space. Memory cards are said to be reliable for up 5 years after their initial use, but DNA-encoded information remains stable and readable for millennia.[5] For purposes of timeless storage, DNA may be dried and then protected from water and oxygen, which gives it a nearly infinite stability.

DNA information storage is not new. It has been around since 1988, and one of the first successful projects came from the J. Craig Venter Institute, a nonprofit genomics research organization with facilities in 3 different US states. These investigators were able to encode 7920 bits into DNA.[6] (Prideful, in a synthetic cell, they encoded their names, 3 literary citations, and the address of an Internet site [Table].)

Newer DNA-synthesizing techniques can alter the way base nucleotide pairs are formed, making it easier to encode information and thereafter read it. As mentioned previously, traditionally base pairs are A-T and G-C (remember that nucleotides are measured in pairs because DNA is usually double-stranded). Thus, the number of base pairs is equal to the number of nucleotides in 1 DNA strand. The problem with using the natural sequence of nucleotide base pairs for information encoding is that the G-C pair can be difficult to subsequently read. Therefore, new techniques use novel base pairs: A-C and G-T, which are easy to manufacture and thereafter interpret. With these 2 new base pairs, one also has a binary code: A-C for 0 and G-T for 1. At present, assembling long stable strands of DNA is difficult, so information needs to be parceled in smaller data blocks of DNA called "oligonucleotides " (by comparison, the human genome contains about 3 billion base pairs, so it is a very long strand and the amount of information that it contains is astonishing).

In a recent experiment, Church et al[7] took one of their own books (nearly 54,000-words-long, including 11 images) and used a computer to convert it into a bit stream (they initially thought about encoding *Moby Dick*). They encoded all of the bits of the book into 159 oligonucleotides, each also containing information as to its general position within the text. The encoded DNA was then amplified by polymerase chain reaction* (PCR), and in this way, its base pairs could be assessed, read, and interpreted (similar techniques were used to map the human genome). During the entire process of writing, amplifying, and reading 5.27 megabits of information, only 10 bit errors occurred, a testament to how incredibly exact this technology is. Church et al were able to store in DNA 600 times more information than was previously possible.

As amazing as this seems, one must add to it the fact that this technique used only in vitro procedures, avoiding the controversies of cloning and live genetic manipulations, and it was 100,000 times cheaper than other previous versions.

Synthetic DNA is exempt from the National Institutes of Health usage guidelines and is available to all with the means to manufacture it. The cost of DNA synthesis drops 12-fold per year compared with that of newer electronic media (1.6-fold per year); thus, it is becoming widely available. For example, synthesizing a strand of DNA containing 100 million base pairs cost US $10,000 in 2001 but only 10 cents today. Synthesizing and reading DNA for information-storage purposes will require 6–8 orders-of magnitude improvement. Although this amount of improvement is significant, it will soon become a reality as handheld DNA sequencers become widely available and inexpensive. As the need to store untold amounts of information becomes more pressing, newer DNA-related technologies will be discovered and become less expensive.

In the supporting data from their article, Church et al[7] also bring up some safety and ethical concerns with regard to their experiment.

They state that the DNA fragments they used to encode their book are "unlikely" to replicate themselves or encode anything else that could be biologically active. They do not discard the possibility that if this DNA were left out in the wild, it could get incorporated into a living organism. This last observation seems unlikely because cells tend to expel DNA that is not theirs. However, what would happen if an organism incorporates this foreign DNA has not even been a matter of speculation. Could a cell produce proteins hitherto unknown?

Will that cell die? It certainly will not help us improve our individual knowledge because our bodies lack mechanisms with which to read this DNA and move its information to our brains. Ninety-eight percent of our DNA is now considered to be "genetic junk" (that is, DNA with no apparent function), so perhaps the day will come when we can use this space to encode into each human cell our history and accumulated knowledge.

All of our knowledge placed into highly resistant and self-replicating cells sent out to space in mini-ships may be the best way to explore the possibility of other civilizations existing far away from ours. Security and defense agencies have also considered DNA storage as a means of encryption. This technique was inspired by the World War II microdot technique of Germany, in which an entire page of information was photographed and reduced to the size of the dot at end of this sentence. DNA microdots can be hidden in general genetic material with their locations known only to those who know the primers marking the beginning and end of their specific DNA segments, which can then be resolved and read with PCR.[8] Therefore, information could cross borders in cells and not be subject to Internet counterespionage, and if the person carrying the information is detained, the site harboring the information would be impossible to detect.

Some say that all science fiction eventually becomes reality, and certainly DNA information storage must have sounded like science fiction just a few years ago. In Frank Herbert's novel *Dune* (Clinton Book Company, 1965), spaceships are able to navigate only because their control systems know at all times the positions of all celestial bodies. This tremendous amount of information is not saved in a computer but rather in mutated humans (the Guild Navigators), each controlling a spaceship. The Navigators can do this because their DNA contains all of the information needed for space travel. If a human being has more than 10 trillion cells, it does not seem far-fetched that his or her DNA could contain all of the information in the universe.

Update

Since I wrote this Perspectives, investigators at the European Bioinformatics Institute have found a new and different way to encode information into DNA. Dr. King's "I Have a Dream" speech, a photo, a PDF of Watson and Crick's seminal article, and all of Shakespeare's sonnets were encoded using it. The new method allows for multiple copies of this special DNA to be accurately manufactured. The authors expect their product to last over 10,000 years if kept dry, cold, and dark. Because storing information in DNA is easier than reading it, they suggest that DNA may be the ideal method for keeping information that does not need to be frequently accessed and thus ideal for libraries and government records. Please see the article in *Nature* by Goldman et al.[9]

* In this situation, PCR is easy to use because the makeup of the DNA strand that needs to be amplified is known. Primers that start the reaction can be easily targeted and specific zones can be chosen and then read.

REFERENCES

1. Seralini GE, Clair E, Mesnage R, et al. **Long term toxicity of Roundup herbicide and a Roundup-tolerant genetically modified maize.** *Food Chem Toxicol* 2012;50:4221–31
2. Zhu Y, He X, Luo Y, et al. **A 90-day feeding study of glyphosatetolerant maize with the G2-aroA gene in Sprague-Dawley rats.** *Food Chem Toxicol* 2012;51:280–87
3. **No Drop at the Top of NFI's Top 10 Most Popular List.** http:// www.aboutseafood.com/press/press-releases/no-drop-top-nfi-stop-10-most-popular-list. Accessed December 7, 2012
4. Church GM, Gao Y, Kosuri S. **Next-generation digital information storage in DNA.** *Science* 2012;337:1628

5. Bonnet J, Colotte M, Coudy D, et al. **Chain and conformation stability of solid-state DNA: implications of room temperature storage.** *Nucleic Acids Res* 2010;38:531–46
6. Gibson DG, Glass JI, Lartigue C, et al. **Creation of bacterial cell controlled by a chemically synthesized genome.** *Science* 2010; 329: 52–56
7. Church GM, Gao Y, Kosuri S. **Next-generation digital information storage in DNA** (supplementary material). http://www.sciencemag. org/content/suppl/2012/08/15/science.1226355.DC1/Church.SM.pdf. Accessed December 7, 2012
8. Clelland CT, Risca V, Bancroft C. **Hidden messages in DNA microdots**. *Nature* 1999;399:533–34
9. Goldman N, Bertone P, Chen S, et al. **Towards practical, high-capacity, low-maintenance information storage in synthesized DNA.** http://www.nature.com/nature/journal/vaop/ncurrent/full/nature11875.html. Accessed February 5, 2012

Works previously stored in DNA[a]

Year of Experiment	Usage Description	Specific Contents
1988	Art	Microvenus image[b]
1998	Text	Text from the Bible, Genesis
2001	Text	Parts from a book by Dickens
2003	Text	Parts from "It's a Small World," the main song of a musical boat ride from Walt Disney[c]
2006	Text	"Tomten" a poem by Viktor Rydberg[d]
2010	Watermark	Watermark of a synthetic genome[e]

a

Adapted from Church et al. Image of the external female genitalia representing female Earth. c Encrypted into *Deinococcus Radiodurans*, a bacterium extremely resistant to inhospitable environments. Information resistant to the effects of a nuclear holocaust could be saved in similar cells. d A true example of "living poetry." For other similar projects, I suggest that you Google "Project Xenotext." e Watermarking a cell designed to contain information may help us keep track of it.

SINGULARITY EVENT

> *We're a crowd, a swarm. We think in groups, travel in armies.*
> *Armies carry the gene for self-destruction. One bomb is never enough, the blur of technology. This is where the oracles plot their wars. Because now comes the introversion. Father Teilhard knew this, the omega point, a leap out of our biology. Ask yourself this question: Do we have to be humans forever?*
>
> Don DeLillo, *Point Omega*[1]

As the universe evolves toward its maximum organized complexity, it is said to reach the Omega Point. "Omega Point" is a term coined by Pierre Teilhard de Chardin to describe the evolution of our universe.[2] A Jesuit who later abandoned the traditional teachings of the Roman Catholic Church, Teilhard de Chardin was a philosopher who also trained as a paleontologist and geologist during the first half of the 20th century. He extrapolated the concept of a spiral galaxy to include the entire universe and out of this forged a unique philosophic viewpoint. His universe was compromised by 2 fundamental forces: tangential or rotational (which he also called matter) and radial or centripetal (also called love). Centripetal forces lead to involution—that is, transforming a state of disorganized complexity into a more organized one. The end result of this involution is the Omega Point or the end of the world as we know it. At this Point, the universe finds itself in a state of organized complexity. From the center of the spiraling universe, mankind serves as a conscious observer or one can also conceive it as each person being the center of his or her own universe, which, as time goes by, becomes more organized.

Reaching the Omega Point may not be possible without possessing the 5 attributes assigned to it by Teilhard de Chardin.

These are pre-existing, personal, transcendent, autonomous, and irreversible. We humans are getting closer to the Point, particularly with the aid of computers and related technology.

The Omega Point is the final step before "Singularity" takes place. Once we achieve (or cross into) Singularity, which will be the first and truly major evolutionary step in mankind, we cease to be humans.[3] In the near future, computers will surpass our collective intellect, and our only way to maintain our place in the universe will be to merge with them. When trans humanists speak about the Omega Point, they refer to the point when our use of science and technology will improve our human state, making conditions such as disability, suffering, disease, aging, and even death a thing of the past.[4]

When I was a young teenager, the first time I became aware of transhumanism was watching a television series called *The Six Million Dollar Man*. In that series, after a crash in an experimental airplane, astronaut Steve Austin was fitted with 2 legs, 1 arm, and an eye, all "bionic" and resulting in superpowers that he used in his new job as a secret agent. The series was very successful, and it was not surprising that NBC decided to create a "bionic" woman (with implants in all 4 extremities, 1 eye, and an ear). This female trans human was not well-accepted by audiences, and the series folded soon thereafter. These 2 cyborgs lacked a true improvement in the way their brains worked, so they were not true examples of Singularity. Transhumanism comprises 2 fundamental changes: the incorporation of technology directly into the brain and/or body (like the 2 previous examples) to improve our functions and performance and/or genetic manipulations to improve biologic processes.

True Singularity may not occur with only 1 of these because creating a super intellegence without the supe rbody to maintain it may not be feasible. Many of those opposed to transhumanism see it as "playing God."[5]

It is interesting to think that it may actually be easier to attain intellectual Singularity than corporal Singularity. Although we know the structure of the human genome, understanding how it works and how to alter its workings favorably may not be feasible in the foreseeable future. For many trans humanists, intellectual Singularity may be as close as 45–50 years away, and it will serve as the gateway to corporal Singularity.

[6] The only thing between unlimited human progress and the way we are now is, paradoxically, our brain and its apparently limited capacity (contained as it is in the cranial bones, it cannot develop more volume and accommodate more than the already present 100 billion neurons and its 100 trillion connections). Through amplification of our native intelligence and/or the addition of artificial intelligence, Singularity can take place and progress becomes fast and unlimited. Unleashed, these "human machines" will work to create new and more powerful, perfect ones.

I think that unfortunately, Singularity will be not democratic and will be available only to those with means to acquire it. Can mankind truly evolve if millions (or billions) are left behind as mere biologic humans? Will we create a dual-tiered social system of super humans and humans even more restrictive than our current social and economic models? The idea of Singularity also reflects the fact that it may happen unexpectedly and that we humans will have trouble understanding what to do with it, creating the opportunity for individuals or groups of individuals to profit from it. An intelligence explosion will cause our current social orders to become disrupted before leading to reorganization and development of different socioeconomic systems (reaching their Omega Point) but not before some chaos takes hold.

A major exponent of Singularity is Ray Kurzweil, an author, scientist, and entrepreneur. He has received honorary doctorates from 17 universities.[7] Kurzweil has been called the Thomas Edison (though that may not be a great thing) of our times, and now in his mature years, his research concentrates on electronic music technology, voice recognition, educational aids, and health supplements, and he even manages a hedge fund.[8] As he gets older, he is understandably preoccupied with death and conceives Singularity as the answer to mortality. Kurzweil bases some of his thinking on the concept of Moore's Law. This law describes the long-term trends in computer hardware and its power.[9] The law is named after Gordon Moore, a cofounder of Intel. Basically, it states that computing power growth is not linear but exponential and that because of this it will become a driving force in technological and social change, something that is already happening (think about how we use our iPhones [Apple, Cupertino, California] to check what we say or where we are going constantly).

A doubling of capacity every 2 years and of performance every 18 months has been noted for all computer related hardware, including transistors, power consumption, storage capacity, network capacity, and so forth. As Moore has stated, this exponential growth can be assumed to continue forever. Kurzweil also believes that the development of new technologies will assure that Moore's Law will not come to an end. Because Moore's law applies to all activities of digital computers and these are the same computers being used to study the human genome, our understanding of it may also follow the principles of that law and allow us to manipulate it more efficiently in the future. Moore's Law, however, does not predict the point at which Singularity will occur.

Nano machines housing intelligent power must be developed and injected or implanted in humans before a super intelligent hybrid being is created. Kurzweil seems to think (following the principles of Moore's Law) that the marriage between artificial and native intelligence will start occurring as early as 2050 (of course, to him this is too late because by that time, he will be dead).

Artificial intelligence is only 1 of several ways to enhance our native one. Others include brain-computer interfaces, biologic manipulation and augmentation, and genetic modifications.

It is unclear if 1 or more of these will be needed to reach Singularity. If human intelligence is the highest we know, it is difficult for us to conceive of intelligence beyond it.

Future computers themselves will be smart enough to build better machines beyond those that any human may conceive.

Neuronal transmission spikes at about 2 Hz per second, and modern computers already spike at 2 gigahertz per second!

Experts are aware that a faster intelligence may not mean a better one. A better brain must make smarter, faster, and self-improving features that are difficult to come by with our current ones and may take centuries to achieve by just normal evolutionary changes. For me, it is difficult to think that with our current intelligence, we will find a cure for cancer or, let's say, Alzheimer disease. Perhaps by creating a super intelligence, these problems will be easier to solve. Increasing intelligence, health, and lifespan are the goals of Singularity, but at this time, it is difficult for our brains to conceive how this utopia will be achieved. To solve our problems as humans, we need to use new tools and not old ones. Einstein said, "The problems that exist in the world today cannot be solved by the level of thinking that created them." Thus, it is also true that with our current levels of technology and knowledge, it is not possible to predict future ones. This trap is quite obvious once one regards the sad state of the World's economy, which we are trying to solve by the old "true and tried" methods of capitalism and free markets.

Conservative thinking may try to impose regulations on changes in intelligence. Just imagine what will happen if the US Food and Drug Administration attempts to regulate Singularity.

Singularity can only occur in free forward-thinking societies, with well-thought regulatory methods and no interference from self-interested parties. If we create laws that block or obstruct Singularity, it will happen in other countries and societies that do not have these restrictions. This is the same principle that we see now with our medical research, which has become so difficult to do in the United States (15 years ago, 20% of articles published in the *American Journal of Neuroradiology* came from outside the United States compared with more than 75% today). Forward-thinking bodies of research that address these issues have been created. The Web site of the Singularity Institute for Artificial Intelligence makes for fascinating reading.[10] In 2009, Kurzweil, among others, helped to establish Singularity University. Physically based at Ames Research Center in California at the National Aeronautics and Space Administration, it was funded by industry leaders such as Google.[11] About 40 individuals serve as teachers at the University, which offers courses costing about US $25,000 (the least expensive, a 10-day "executive" program, costs US $15,000). For the first course in 2009, the University received more than 1200 applications from which only 40 were selected.

Their current Web site states that 4 yearly selections occur, resulting in an acceptance rate of 25%. They offer courses in the following tracks: technology (which includes biotechnology, medicine, and neuroscience among others), resources, and applications. Each track follows a similar class schedule: week 1, understanding the field; week 2, learning about exponential growth; and week 3, actionable output.[12] Of the faculty, only Christopher de Charms seems to have some relationship with neuroimaging (in his case, functional MR imaging) among those listed in the Medicine and Neuroscience curriculum. The part of the curriculum directly related to imaging gives the following description, "Medical diagnostics and imaging: increasingly powerful and rapid imaging modalities, point-of-care medical diagnostics, nano medicine and biomarker technology."[13] To someone like myself, an academician educated in public hospitals and traditional university structures, their ideas sound a bit commercial and certainly make me wonder about conflicts of interest (how can you earnestly teach something when you own stock in companies that produce it?), but maybe I am being too old-fashioned.

It is becoming clear that radiology and interpretation of imaging studies will be altered by new forms of intelligence. In February 2011, an IBM computer named Watson beat several previous champions at *Jeopardy,* demonstrating that artificial intelligence is no longer a thing of the future. Watson is capable of understanding the nuances of spoken English and answers faster (and better) than humans. So, in a mostly visual specialty like radiology, a machine could be much better at analyzing the images and pinpointing the abnormalities. The industry is already starting to think about developing such machines for this purpose. Very soon we will have to incorporate millions of individuals into our existing health system and utilize our imaging equipment more efficiently. It is clear that there will be a significant lack of radiologists, resulting in a very complex situation. So first, we need to reach our Omega Point and organize the complexity of our specialty. Then we could have all studies screened by a computer and just look at the abnormal ones. The last step would be to achieve Singularity with one of these computers and still be radiologists, only better and faster ones.

References

1. DeLillo D. *Point Omega.* New York: Scribner; 2010:53
2. Teilhard de Chardin P. *The Phenomenon of Man.* New York: Harper Collins; 2002
3. Technological singularity. Wikipedia. http://en.wikipedia.org/wiki/Technologi cal_singularity. Accessed April 1, 2011 **394** Editorials _ AJNR 33 _ Mar 2012 _ www.ajnr.org
4. A history of transhumanist thought. www.nickbostrom.com/papers/history. pdf. Accessed April 1, 2011
5. Transhumanism: the world's most dangerous idea? http://www.nickbostrom. com/papers/dangerous.html. Accessed April 1, 2011
6. The Singularity Is Near. Wikipedia. http://en.wikipedia.org/wiki/The_Singu larity_Is_Near. Accessed April 1, 2011
7. Ray Kurzweil. Wikipedia. http://en.wikipedia.org/wiki/Ray_Kurzweil. Accessed April 1, 2011
8. Kurzweil Technologies. http://www.kurzweiltech.com/ktiflash.html. Accessed April 1, 2011
9. 1965–"Moore's Law" predicts the future of integrated circuits. Computer history museum. http://www.computerhistory.org/semiconductor/timeline/1965-Moore.html. Accessed April 1, 2011
10. Singularity Institute. http://singinst.org. Accessed April 1, 2011
11. **Merely Human? That's So Yesterday.** *New York Times.* Business Day. June 12, 2010. http://www.nytimes.com/2010/06/13/business/13sing. html?pagewanted _2&sq_singularity%20university&st_cse&scp_1. Accessed April 1, 2011
12. Are you ready for the future? Singularity University. http://singularityu.org/programs/graduate-studies-program/#admissions. Accessed April 1, 2011
13. Future studies and forecasting. Singularity University. http://singularityu.org

Science of brain imaging:

IMAGING IN SPACE EXPLORATION

I remember the amazement I felt when man walked on the moon. Today, I feel that monies spent on space exploration would be better used in green energy and ocean preservation programs. We already live on earth, the most perfect spaceship of all: Why go beyond it? Regardless of my feelings, space exploration will continue, funded either by governments or private industry. For example, for only US $200,000 (a deposit of $20,000 is required), you can book a suborbital space trip on Virgin Galactic; 340 places are already reserved and paid for.[1]

If you want to experience microgravity cheaper, you can go to Las Vegas and fly the "Vomit Comet." If you desire a longer journey, Space Adventures will take you on a 10-day trip to the International Space Station (ISS) for US $25 million, and they even offer flights to the moon.[2] SPACEX is a private company that is nearing completion of its human space transporter, one that NASA will probably use once the Space Shuttle program is grounded.[3] Other private companies developing transporters include Armadillo Aerospace, XCor Lynx, and Blue Origin (owned by Jeff Bezos who is CEO of Amazon). These transporters may also be used for space tourism and for research.

So what happens if we get sick in space? Techniques for conducting physical examinations in microgravity have been studied.[4] These experiments have been done during parabolic flights, space shuttle missions, and longer sojourns at the ISS.

During these trips, some gravity is still present because most happen close to earth (micro- or partial gravity environment).

[5] For example, in the ISS, gravity is about 88% of that felt at ground level here on earth (astronauts seem to be floating due to the fact that they are traveling at about 17,500 miles per hour). In 1 study, physician-astronauts were asked to evaluate the effects of microgravity on the cardiovascular, musculoskeletal, and neurosensory systems. The evaluation of the latter included only reflexes (which were brisker in space).

Head and neck radiologists may be interested in the fact that facial and nasal mucosal swelling are common during space travel (due to fluid redistribution).* The normal flexed position the human body assumes in space may have effects on the spine, and we know that astronauts are taller when they come back from their missions. When arriving back from Mars, it is estimated that bones will have lost nearly 60% of their attenuation (it takes nearly 1 year to recover bone attenuation lost during 1 month in space), so astronauts will be at increased risk for vertebral fractures. Some think that complete bone mass recovery after prolonged space trips is not possible.

Decompression sickness is not uncommon in space and, as we all know, it may affect the brain. Inside a space suit, "atmospheric air" is purged of all nitrogen and replaced completely by oxygen. The pressure inside the suit is about one-third of normal to allow easier breathing and motion. Before an astronaut dons the suit, he or she must go through decompression to eliminate circulating nitrogen, and this is also done when coming out of it. During depressurization, air expands and may result in ear and sinus pain, decreased hearing, and mandible and tooth pain. The inner ear is affected mostly during the first 2 days of a space trip and triggers loss of balance and dizziness. Dizziness and fainting also are caused by orthostatic intolerance. An astronaut's blood volume is reduced by 22%, affecting cerebral blood flow. Sleep disturbances are very common; most space travelers get only about 2 hours of sleep and, to get any, they must take strong sedatives. Alertness is reduced and performance errors are more common, which is typical if you are sleep-deprived. Feelings of isolation, depression, and other emotional disturbances are also typical.

As extraterrestrial trips last longer, 2 critical situations that may affect our nervous systems arise: immune system weakness and exposure to radiation. T-cells do not function normally in space, making individuals more susceptible to infections.

Valentin Lebedev, a cosmonaut, spent 221 days in orbit and later lost his eyesight due to cataracts as a consequence of radiation exposure. In one study, 48 cataracts developed in 295 astronauts, particularly those who were on longer missions in Skylab, the Space Shuttle, and the ISS.[6] The cause was probably genetic damage in epithelial cells, disruptions of cell cycles, apoptosis, abnormal differentiation, and cellular disorganization due to exposure to high-energy protons and heavy ions. Because eyesight problems account for 40% of disqualifications in astronaut selection, cumulative radiation exposure plays an important role in recurring space trips. Radiation, particularly cosmic radiation, easily penetrates aluminum, the most commonly used material in space ships.

The International Commission on Radiation Protection sets a limit of 20 mSv/year for commercial flight crews.[7] The highest doses of radiation (about 5 mSv/year) in commercial aviation have been registered for Concorde crews. Cosmic radiation becomes a problem with longer missions, and going to Mars will take about 2.5–3 years.

To control radiation exposure, new building materials are being designed and strict monitoring of exposure levels will be needed. It has been estimated that middle-aged astronauts will probably do well as the effects of this exposure will not be evident during their normal life span and probably they are past the age of having children.[8] NASA has set a limit of 6 Sv lifetime skin exposure (though it varies somewhat with age and sex and for specific organs). Exposure of more than 1

Sv/year induces cancer. Exposure to 2–4 Sv/year results in chronic radiation syndrome with complex clinical symptoms.

Because cosmic radiation comes in great part from the sun, living in regions exposed to maximum solar activity (such as the moon's South Pole, a proposed place for a refueling station on our way to Mars) may result in high acute single doses. A single exposure between 3 and 5 Sv will most likely kill you due to bone marrow and gastrointestinal syndromes.[9] It may not take long to die from the gastrointestinal syndrome if you vomit inside your space suit (though the ventilation systems in space helmets are designed with this in mind).

Most of these disorders will take place once individuals return to earth, so diagnosing and treating them in the usual fashion will not be a problem as it would in space. However, with space tourism and prolonged space journeys just around the corner, we may need to diagnose and treat some problems in space. For example, under normal gravity conditions, urine exerts pressure on the bladder floor resulting in the need to urinate, but in zero gravity, overfilling results in urethral compression and urinary retention. This why astronauts urinate on a "preventive schedule" and know how to catheterize themselves.

Tele assistance in medical conditions leads to 2 situations: evacuate or treat on site. Companies providing telemedicine services to remote sites such as Antarctica and Mount Everest have the option to order evacuation, but in space, we will probably have to treat there (though astronauts in low-altitude missions have been deorbited when sick). Transmitted data now routinely include temperature, pulse oxymetry, electronic stethoscope, graphic files, and videoconferencing. Devices using near-infrared spectroscopy that are capable of monitoring a variety of physiologic parameters are being developed.

[10] In space exploration, transmit lags need to be kept in mind (40 minutes from Mars to earth, so any real-time interactions are not possible).

Still, we have learned from the experiences of 23 US physicians who have traveled in space that some treatments up there may be possible. Space health monitoring started in 1961 with Yuri Gagarin (and before that with monkeys). Here we need to remember that at the start of the space programs, we did not know the answers to simple questions: Can the heart beat in space? Can we urinate in space? Can we swallow in space? and so forth. As spaceships became bigger, the possibility to diagnose and treat disorders in space became a reality. The US and Russian space programs estimate 1 emergency in space per year and the requirement for advanced life support and/or anesthesia at once every 3–4 years.[11] For cardiopulmonary resuscitation, several methods of chest compression have been attempted, but none work well in micro- or partial gravity.

Intubation has been studied, but it probably takes too long to be of any good. Surgery has only been tried in animals, and collection of any organic materials released is a (big) problem, though drawing blood is feasible and safe. Overall, space crews receive about 40–60 hours of medical training, and, when combined with telemedicine, the quality of care is comparable to that available in an earthbound ambulance.

Evacuations would require a permanently available vehicle or a second constantly orbiting one. Limited space in these vehicles is not optimal for procedures when evacuating an acutely sick individual. Emergency ballistic re-entries can exceed 7*g*, and though humans can perform well in hypergravity conditions, devices such as Ambu bags (Ambu, Copenhagen, Denmark) do not. Radio communications are not possible during re-entry. Nevertheless, evacuations from even the moon may be faster than those from Antarctica during winter.

In the 1980s, Soviet cosmonauts performed sonography in space, and in 2004, sonography was used to evaluate the shoulders of astronauts in the ISS. Shoulder evaluation may be needed after strenuous extravehicular activities when astronauts complain of pain. Astronauts received a 5-hour course 4 months before launch and completed a 1-hour enhancement program once onboard the ISS.[12] Examinations took only about 15 minutes, and, when combined with teleguidance, were thought be good enough for decision-making. Sonography in space has also been used to evaluate the heart and the intervertebral disks.[13] Sonography may not be adequate for evaluation of the central nervous system, though it could potentially diagnose abnormal flow in major arteries and then, with focused applications, could be used to enhance fragmenting of clots in patients with stroke.[14]

A NASA project deals with the stability of medications in space.[15] Radiation may alter chemical stability and diminish potency and shelf life, something particularly true of vitamins and amino acids that are essential to maintain a healthy diet on long trips. These experiments may lead to alternate manufacturing, storage, and dispensing methods.

Ideally, larger spaceships could carry CT or even MR imaging scanners. Several teams of investigators are developing CT units that use electron beams guided by magnets and thus have no moving parts. These units could potentially become small enough to be deployed to combat zones and later into space. Optical coherence tomography is being tested in catheters and endoscopes and theoretically could be applied to sonography, CT, and MR imaging. Hand-held MR imaging units may one day be possible as superconducting quantum interference devices improve. A German laboratory has developed a palm-sized magnet capable of 0.7T.[16] A hand-held sonography device that looks almost exactly like an apparatus Dr McCoy used in the *Star Trek* series is available.[17]

In reality, I think that there will be very little, if any, use for neuroimaging in space. If images are generated, we certainly can interpret them down here on earth. That's what I like to think of as the ultimate moonlighting . . . or perhaps "Marslighting."

* This is part of the so-called NASA beauty treatment: redistribution of abdominal organs results in a smaller waistline, facial edema erases wrinkles, and appendages sag less with age. For more about this and other space travel issues, I recommend Mary Roach's *Packing for Mars: The Curious Science of Life in the Void*. New York: WW Norton, 2010.

References

1. Virgin Galactic. http://www.virgingalactic.com. Accessed March 3, 2011
2. Space Adventures. http://www.spaceadventures.com. Accessed March 3, 2011
3. SPACEX. http://www.spacex.com. Accessed March 3, 2011
4. Harris BA, Billica RD, Bishop SL, et al. **Physical examination during space flight.** *Mayo Clin Proc* 1997;2:301–08
5. NASA.GlenResearchCenter.http://www.nasa.gov/centers/glenn/shuttlestation/ station/microgex.html. Accessed March 3, 2011
6. Cucinotta FA, Manuel FK, Jones J, et al. **Space radiation and cataract in astronauts.** *Radiat Res* 2001;156:460–66

7. Health Physics Society. http://www.hps.org/publicinformation/ate/faqs/commercialflights.html. Accessed March 3, 2011
8. Simonsen L, Wilson JW, Kim MH, et al. **Radiation exposure for human Mars exploration.** *Health Phys* 2000;79:515–25
9. Hellweg CE, Baumstark-Khan C. **Getting ready for the manned mission to Mars: the astronaut's risk from space radiation.** *Naturwissenschaften* 2007;94: 517–26. Epub 2007 Jan 19
10. Biomedical Engineering and Medical Physics. Babs Soller, PhD. http:// www. umassmed.edu/biomedeng/faculty/soller.cfm?start_0. Accessed March 3, 2011
11. Cermack M. **Monitoring and telemedicine support in remote environments and in human space flight.** *Br J Anaesth* 2006;97:107–14. Epub 2006 May 26
12. Fincke EM, Padalka G, Lee D, et al. **Evaluation of shoulder integrity in space: first report of musculoskeletal US on the International Space Station.** *Radiology* 2005;234:319–22
13. Martin DS, South DA, Garcia KM, et al. **Ultrasound in space.** *Ultrasound Med Biol* 2003;29:1–12
14. Alexandrov AV. **Ultrasound enhancement of fibrinolysis**. *Stroke* 2009;40(3 suppl):S107–10. Epub 2008 Dec 8
15. NASA. International Space Station. http://www.nasa.gov/mission_pages/station/science/experiments/Stability.html#applications. Accessed March 3, 2011
16. Technology Review. Published by MIT. Palm-sizeNMR.http://www.technology review.com/biomedicine/25527. Accessed March 3, 2011
17. GE. http://www.ge.com/audio_video/ge/health/meet_vscan.html. Accessed March 3, 2011

THE INDUSTRY OF CT SCANNING

Let's recap: CT was invented in 1972 by Godfrey Hounsfield of the EMI Laboratories in England and by Allan Cormack of Tufts

University in the United States. In 1979, both shared the Nobel Prize for its invention. The first commercial units (1974) were for head imaging only and acquired data only 1 section at a time, which, in turn, took hours to reconstruct.

Two years later, whole-body scanners became available. In 2007, 72 million CT scans were done just in the United States.[1]

Noone knows exactly how many machines have been sold, but numbers vary between 6000 and 7000 in the United States (24–25 per million population) and probably more than 30,000 worldwide. The main limitations of medical CT are that the entire apparatus needs to rotate around the patient, resulting in complex mechanics and frequent calibrations, limited spatial resolution, and radiation exposure. Recently, spatial resolution has been hampered by increased noise due to lower radiation doses. Although when we think CT, we think radiology and medicine, its applications are more far reaching and the business of scanning is no longer limited to medicine. Most of the limitations of medical CT are not shared by industrial CT as we will see. In this *Perspectives*, I will discuss some applications of CT scanning outside of medical imaging.

The major industrial applications of CT include research,inspections, attenuation analysis, reverse engineering, measurements, and 3D digitization. The industry refers to inspections using CT as "NDT" (nondestructive testing), and there are many companies offering this service. (There is even an American Society of Non-Destructive Testing [http://www. asnt.org]). NDT is essential when lightweight materials such as aluminum are used for high-stress-resistant structures like honeycombs. Honeycombs offer great structural integrity at a fraction of weight compared with other materials, but with stress, they may fracture (generally hairline-type fractures).

Because honeycombs are used in essential parts of airplanes (engines, exhausts), they need to be checked regularly, and CT scanning is a great way to do this. NDT is also used for examinations of gasoline-burning motor castings and plastic and composite materials. Molds can be scanned, and information obtained may be used to assess integrity and for duplication.

Apart from aerospace, aviation, and automotive applications, industrial CT may be used to inspect the inside of electronic equipment such as radios, and because of its extreme spatial resolution, some machines are capable of imaging the inside of computer chips and cables that conduct electricity or optical fibers. In museums, CT scanners are commonly used for the examination of skeletal remains, mummies, vases, and other artifacts. In dental science, industrial micro-CT allows 3D visualization and extremely accurate measurements of dental implants and prostheses. Pharmaceutical companies use it to image the inside of tablets and capsules and to assess the internal components of different apparatuses such as asthma inhalers. Foods may be inspected, and one commonly sees this application in airports where the Department of Agriculture scans luggage looking for food and then scans the products that are uncovered. Metrologists (scientists specializing in the science of measurements) use CT scanning routinely.

Industrial micro-CT is a different ballgame than human CT scanning. One rotation may take up to 1 hour in duration but results in images with resolutions between 1 and 10 _m, depending on the system (generally their resolution is more than 100 times better than that found in the best medical CT units; remember that it takes 10,000 _mto make 1 cm). Total scanning time is generally between 45 minutes to 5 hours, though newer generations of these machines can scan simple objects in a matter of seconds. The part to be scanned is placed inside a chamber on a rotating platform (in some scanners, parts that have been scanned are removed and exchanged for others by using robotic arms, which is faster than doing it manually). This is one of the main differences between industrial and medical CT scanning. In the former, the object rotates; in the latter, the x-ray equipment does. Rotating the x-ray tube and detectors limits acquisition speed and leads to loss of calibration and the need for frequent adjustments.

In industrial CT units, the scanning chamber size varies, but it tends to be relatively small compared with what we radiologists are used to (in our scanners, the gantry opening is the equivalent of this chamber). Because industrial materials may be very dense, these machines use significant doses of radiation (450 kV is typical), and megavolt units are capable of penetrating up to 400 mm of steel. Typical industrial scans produce anywhere between 360 and 3600 images "weighing" about 2 GB in total. Industrial scanning companies also offer 3D reformations, and their computers are capable of processing billions of voxels in just a few seconds.[2] The price for each of these systems varies between US $300,000 and $5 million. I was not able to find what an individual study costs as most companies require a consultation before giving you a price estimate.

For many years, industry has used laser scanners to obtain exact measurements, but these offer only external evaluations, whereas CT is capable of external and internal analysis. This has lead to industrial CT being extensively used in the field of reverse engineering, which is the process of discovering the technologies used by humans to build things. Reverse engineering is commonly used in industrial espionage, and the military use it extensively to study and copy products of other countries. One of the most intriguing applications is reverse engineering of the brain. Developers of artificial intelligence pay close attention to the way that our brains (or those of animals) are engineered and try to copy parts of these. It is hoped that this activity will extend to the physiologic activities of the brain and that models of brain-medication interactions will be designed and studied,[3] resulting in better treatments.

Reverse engineering of the human body has lead to the development of improved extremity prostheses, artificial retinas, and other neural prostheses (such as cochlear implants).

The other area where industrial CT is commonly used is rapid prototyping (solid free-form fabrication). This refers to the use of CT data in the construction of physical objects (models and prototypes). When consecutive layers of materials are added to make objects from 3D data, one is said to be doing additive manufacturing; the reverse is called subtractive manufacturing. Rapid prototyping is also used by artists, sculptors, and jewelers to produce complex shapes. It is expected that in the not-so-distant future, rapid prototyping will allow one to create objects at home. In the world of nanotechnology, rapid prototyping is used for structure design, simulations, interfaces, patterning, and other manipulations from simple molecules at the atomic scale to complex nano devices.

Nano-CT scanners are a new type of device with spatial resolutions of around 400 nm. One manufacturer claims to be able to image individual osteocytes in a mouse fibula. Carbon nanotube CT scanners will probably be cheap, work at room temperatures, be very fast, have no moving parts, and be of different sizes, some small enough to be deployed to war theaters and space.[4]

Contrary to popular belief, airports do not routinely use CT equipment to scan luggage. Most luggage scanners are simple x-ray machines that assign different colors (generally 6) to objects according to their atomic numbers (you may remember from your physics courses that the atomic number is the number of protons in the nucleus of an atom). The so-called CTX scanners are used specifically to discover explosives.

These units are very similar to ours—that is, the luggage remains static in a gantry and an x-ray tube rotates around it at 120 revolutions per minute (in the fastest of these machines), generating an image every 0.5 seconds and scanning up to 560 bags per hour. When CT scanning becomes available at the passenger security checkpoints, 3D reformations may allow identification and internal examination of portable computers, and these should no longer need to come out of our bags.

CT scans of passenger luggage are essential when a person gets "flagged" as high risk.

When one's body is scanned at the airport, images are obtained with "backscatter" (narrow and low-attenuation beams) or radio-frequency energy (millimeter wave) or with whole-body "advanced imaging technique" (AIT) scanners that are very similar to conventional CT scanners.[5] Both scanners create a 3D surface image of the body that some believe violates individual privacy (though genitals are hidden from view). Studies have found that backscatter scanning is safe for the general public, even for children who are more sensitive to radiation exposure than adults.[6] These scanners use a very low-level millimeter-wave nonionizing radiation that does not penetrate the skin (there is debate if this type of radiation is harmful to individuals predisposed to skin cancers, and because most of the radiation strikes the head where 85% of basal cell cancers occur, the risk may be real). The dose received is equivalent to that of flying 2–3 minutes at 30,000 feet on a commercial airplane according to the American College of Radiology.[7] Between 1000 and 2000 of these scans are needed to reach the radiation level of a chest radiograph.[8]

AIT scanners do use radiation, and though the dose they deliver is safe per individual, it may not be safe to the population as a whole. Experts estimate that up to 100 new cancers per year may occur secondary to radiation exposure at airport security posts in the United States alone. Because radiation is absorbed by the skin, children may be more sensitive to its effects than adults. What this radiation exposure will mean to unborn children is the concern of many mothers.[9] Although the total time needed to scan each passenger is between 8 and 15 seconds, radiation is delivered during only a fraction of a second. The long-term effects of being repeatedly exposed to low-level radiation (which would a problem for pilots who are scanned hundreds of times each year and additionally get the exposure from high-altitude flights) has not yet been ascertained.

Curiously, accidental overdoses need not be reported.

Each of these machines costs about US $170,000 and as of this writing, there were 400 in 69 airports. A cost-benefit study of these machines has not been performed as the Transportation Safety Agency (TSA) states that the Department of Human Services does not require it. The cost of these scanners is staggering, about $336 million from the over $8 billion TSA budget. *USA Today* reported that airport scanner manufacturers doubled their lobbying expenditure in the last 5 years with a single company spending over $4.5 million.[10] As a traveler, you may refuse a body scan but not a full-body pat down in its place (to do so means that you will not be allowed to fly and may be fined $10,000). To put the quantity of airport scans needed to ensure air safety in perspective, remember that in 2010 more than 90 million passengers went through just the Atlanta airport (much more than the number of patients who got medical CT scans during the same period of time). It is expected that soon, 1 billion passengers will be scanned in the United States every year, and with exposure such as this to the population, there is a real risk for harm. In *America On-line News*, the TSA stated that the US Army, the US Food and Drug Administration (FDA), and The Johns Hopkins University were involved in assessing the safety of these scanners. Surprisingly, all 3 institutions deny this claim. Because airport scanners are not considered "medical," the FDA has no jurisdiction over them. The Health Physics Society has requested radiation data from TSA, only to have it denied several times.

Congress has now called on the TSA to release these data for independent analysis but that has not occurred at the time of this writing. This issue becomes critical because new airport scanners that are able to do "cavity searches" are being developed, and these use penetrating radiation very similar to that used in medical CT scanners.

So, now you may stop wondering about why your CT salesperson does not visit you anymore or why these companies do not support our scientific meetings the way that they used to.

The US medical CT market is basically saturated and that may soon happen in most other developed countries; why not go outside of medicine? Our regulatory agencies— quite adequately— are keeping a close eye on medical CT due to the recent radiation overexposure situations; why not go outside of medicine where radiation exposure is of no concern? Airports and maritime ports will soon be buying more scanners, but they are not obligated to report, justify, or explain their use (or abuse) to anyone as we are in medicine. The business of CT may not be growing in your practice, but you can rest assured that it is booming elsewhere.

References

1. Berrington de Gonzalez A, Mahesh M, Kim KP, et al. **Projected cancer risks from computed tomographic scans performed in the United States in 2007.** *Arch Intern Med* 2009;169:2071–77
2. **Advantages of CT in 3D scanning of industrial parts.** http://www.3dscanningtechnologies.com/pdfs/parts.pdf. Accessed May 20, 2011
3. **Grand challenges for engineering. Reverse-engineer the brain.** http://www. engineeringchallenges.org/cms/8996/9109.aspx. Accessed May 20, 2011
4. www.physics.unc.edu/project/zhou/papers/pdf/gcao_spie_2010.pdf. Accessed May 20, 2011
5. Brenner DJ. **Are x-ray backscatter scanners safe for airport passenger screening? For most individuals probably yes, but a billion scans per year raises long-term public health concerns.** *Radiology* 2011;259:6–10 **584** Editorial _ AJNR 33 _ Apr 2012 _ www.ajnr.org
6. Frisman P. **Full body scanners. OLR research report.** December 7, 2010. http:// www.cga.ct.gov/2010/rpt/2010-R-0494.htm. Accessed May 20, 2011
7. American College of Radiology. **ACR statement on airport full-body scanners and radiation.** http://www.acr.org/MainMenuCategories/media_room/Featured Categories/PressReleases/StatementonAirportFullbody Scanners.aspx. Accessed May 20, 2011
8. www.medscape.com/viewarticle/735717. Accessed May 20, 2011
9. Worthington A. **Airport travelers to get ionizing x-ray radiation.** http://www. rense.com/general41/airporttravelerstoget.htm. Accessed May 20, 2011
10. Schouter F. **Body scanner makers doubled lobbying cash over 5 years**. *USA Today*. November 11, 2010 http://www.usatoday.com/news/washington/2010–11-22-scanner-lobby_N.htm. Accessed May 20, 2011

STRONG MAGNETS

As radiologists, we are familiar with the use of magnets to generate images that are now a mainstay of radiology.MR imaging is still evolving, with ever increasing magnetic field strengths. Nowadays, 3T magnets are commonplace, with 7-9T research systems in existence and 11T human systems under development. Most of us know that smaller bore magnets, capable of higher field strengths, are also used for MR imaging on samples ranging from humans to animals and isolated tissues down to single cells. MR spectroscopy itself, used mainly for elucidating the chemical content of materials and solutions, continues to find applications in biological research and human studies. As neuroradiologists we forget that MR imaging is only 1 part of an extensive range of applications for magnets, and that at facilities throughout the world, an astounding variety of magnet systems are being developed and applied for a wide range of uses outside the realm of clinical medicine. In this perspective, I give a short overview of some of these facilities.

Tallahassee, the capital of Florida, is located in the center of that states' panhandle, just south of its border with Georgia. In a city of about 200,000 inhabitants, between its regional airport and downtown, one finds the National High Magnet Field Laboratory (or, for short: Mag Lab). This facility comprises a group of angular, clean, bright buildings (covering over 370,000 sq ft), making it the largest facility of its kind in the world. The Mag Lab attracts about 1000 visiting scientists every year. Before the Mag Lab, the largest and most prestigious laboratory of this sort was at the Massachusetts Institute of Technology (MIT). The Boston lab was named the Francis Bitter Laboratory in 1967, and in the 1970s began its operations and obtained funding from the National Science Foundation.

Today, it houses several 17.6T magnets and a 21T unit, all dedicated to research.[1] During the 1980s, 3 institutions—

Florida State University (FSU), Los Alamos National Laboratory (Pulsed Field Facility), and the University of Florida— proposed the creation of Mag Lab to the National Science Foundation. To the surprise of the folks at MIT, the state of Florida was awarded the right to build what is today the most complete research ultra-high-field MR facility in the world.

Completed on time, the center was dedicated by former Vice President Al Gore in June 1994. Mag Lab research includes physics, biology, bioengineering, chemistry, geochemistry, biochemistry, materials science, and engineering. It also houses the Center for Research and Learning, which serves as its education arm.

When Mag Lab opened, it housed a 27T magnet, at that time the strongest in the world. One year later, a 30T resistive magnet rivaling hybrid systems found at MIT and the Laboratoire National des Champs Magnetiques Intenses in Grenoble, France, was ramped up. In France, powerful research magnets producing continuous fields are housed in Grenoble, whereas those producing pulsed and even higher strengths are housed in the Toulouse facility (see below).[2] Similarly, the Mag Lab facility houses the continuous magnets in Tallahassee, while the pulsed systems are at Los Alamos. By 1998, Mag Lab total investment in infrastructure from all sources was US $192 million, including the creation of a resistive magnet to be deployed to the International Space Station, in addition to outside contracts such as building a 30T unit in Tsukuba, Japan, which is the highest field resistive magnet in Asia.

In 1999, Mag Lab made it into the *Guinness Book of World Records* when it built a continuous field 45T hybrid MR system.

Pictures of this apparatus show it to resemble some infernal machine from Nikola Tesla's imagination (such as the ones built during his Colorado years) or, perhaps, the core of a nuclear reactor. Several stories high, this vertical magnet has a bore opening about the size of a golf ball (3.2 cm). Tiny samples of the materials to be interrogated descend into the bore via a probe. To reach such enormous magnetic strength, the magnet is of hybrid design—that is, resistive and superconductive at the same time. So in reality, it is 2 magnets in 1. The inner magnet is a resistive one surrounded by a superconductive one.

One can request magnet time by sending in a short 3-page description of a project, and a committee will decide which magnet and which facility is most appropriate for each individual experiment. Generally, researchers are granted 1-weeklong periods, so they work around the clock to complete their projects during those short 7 days. Although the larger magnets are expensive to run (about US $4000–5000 per hour) magnet time is paid for by Mag Lab and is thus made available for users free of charge. Additional support such as computers, cryogenics, an electrical shop, and so forth is also available. A list of visiting scientists (including names, parent institutions, and title of project) is available on-line.[3] At the Tallahassee site most work is in the materials sciences with personnel from chemistry, physics, astronomy, or engineering departments.

Most biology and medicine studies are performed at the University of Florida site. This last site also houses a Mag Lab supported high B/T facility, which conducts experiments in strong magnets at very low temperatures close to absolute zero (_273 Kelvin).

An interesting series of lectures is held at Mag Lab throughout the year, and the program and a short explanation of it are also found on their Web site. Among the different topics, one can find a vodcast by Sir Harry Kroto, a Nobel Prize winner for chemistry, chatting about great minds of the 16th and 17th centuries. Most interesting, he states that we are not in the midst of a scientific revolution but rather of a technical one.

Mag Lab has received many accolades and does not rest on its laurels; this year they completed construction of the highest field strength (36.2T) resistive magnet in the world.

The Magnet Lab at Los Alamos National Laboratory also houses impressive equipment. Pulsed magnets there are capable of generating fields of up to 100T for very short periods of time. A single turn magnet can generate up to 300T for a 6-_s burst but is destroyed by explosives in the process of doing this. For years, magnet engineers have thought of 100T as the Holy Grail for non-destructive magnets. Materials used at these field strengths have enormous tensile strength since the strong magnetic fields result in energies equivalent to 200 sticks of dynamite.[4] Eight other magnets, the weakest being 17T, are currently functioning at Los Alamos. The facility looks exactly like what one expects to find in Los Alamos: boxy, brown, secretive-looking, nondescript buildings. Al- though nationally sensitive work there is off limits to the public, the Mag Lab part of the facility is open and accessible. Los Alamos National Laboratory is located in New Mexico, and the largest nearby town is White Rock; the better known Santa Fe is about 35 miles away. The entire Los Alamos site is a group of about 1800 buildings spanning 35 square miles.

The third and complementary site of this effort resides at the University of Florida in Gainesville, a city of about 130,000 inhabitants located in northern Florida about 100 miles from Disney World. The University is quite large with over 51,000 students and nearly 5000 faculty members. There the Mag Lab facility is located in the McKnight Brain Institute and involves MRI and MR spectroscopy, hence its name: the Advanced MRI and Spectroscopy (AMRIS) facility. The Institute is one of the world's biggest neuroscience research operations with a faculty of over 300. There one finds 7 magnets, including an 11T unit with a 40-cm bore and a whole-body 3T unit. Unfortunately, their Web site seems a bit anachronistic and simple and does not offer a great deal of information.

The Laboratoire National des Champs Magnetiques Intenses in Grenoble is one of the main institutions belonging to the Centre National de la Recherche Scientifique in France and is open to researchers from the 27 states of the European Union and adjacent countries such as Turkey, Israel, and others.

Grenoble is located in southeast France, close to the Italian border and at the foot of the Alps, a location that has earned its nickname: the Capital of the Alps. Housed in this laboratory is a 35T magnet with a 34-mm-wide bore The Laboratoire National des Champs Magnetiques Intenses in Toulouse is found in southwest France. Because Toulouse is also known as La Ville Rose, it is not unexpected that its building has pinkish tones. There one finds magnets capable of 45T or 60T during pulses as long as 1 second in duration and of 150T–260T for only microseconds. Both of these facilities form part of the larger EuroMagnet Net II (Research Infrastructures for High Magnetic Field in Europe).[5]

Other facilities that are part of EuroMagnet Net II include the High Field Magnet Laboratory in Nijmegen, the Netherlands.

[6] This laboratory is housed in a beautiful modernist building that has a curvy, sensuous facade and contains 32T magnets and is building a new 32 mm magnet which they claim will make 38 T and a hybrid one which will make 45 T.

As well as an extensive magnet research and application program in many ways similar to the Mag Lab, some of their work applies magnetic levitation such as is used in magnetic levitation trains and levitation displays (used in those globes that seem to float on air). A provocative idea is that humans, if placed in a strong enough magnetic field, can also levitate.

Another EuroMagnet Net II facility is the Dresden High Magnetic Field Laboratory located in Germany.[7] This facility is located in the countryside outside the city of Dresden. It is a part of a large physics campus called the Helmholtz-Zentrum Dresden-Rossendorf. Magnets 70T and above are housed in a no-nonsense, industrial-looking modern building. Laser beams allow spectroscopy at very high field strengths.

The largest similar installation in Asia is the Tsukuba Magnet Laboratories in Japan.[6] This facility was established in 1993, and today it houses 17 high-field strength magnets including resistives, hybrids, pulsed and superconducting magnets.

A superconducting unit capable of producing 24T just became operational. The facility was not open to external researchers until 1998. It is under the direction of the National Institute of Materials Science, which has established a collaborative research effort with the University of Washington in Seattle.

I hope that this short editorial complements my previous one about the industry of CT scanning. It is important for us, clinical neuroradiologists, to realize that magnets are used by other researchers whose areas of interest are very different from ours. I wish to thank Dr Robert Quencer, who gave me the idea for this *Perspectives.*

Acknowledgments

I sincerely thank J.S. Blackband, G. Boebinger, and M. Bird from the Mag Lab Tallahassee and Gainesville locations for their corrections, suggestions, and encouragement.

References

1. Francis Bitter Magnet Laboratory. Current News. http://web.mit.edu/fbml/ index.shtml. Accessed June 17, 2011
2. Laboratoire National des Champs Magne´tiques Intenses. http://lncmi.cnrs.fr/.
   Accessed June 17, 2011
3. National High Magnetic Field Laboratory. The Magnet Lab Media Center. http://www.magnet.fsu.edu/mediacenter/visitors/index.aspx?visitor_0. Accessed June 17, 2011
4. Tsukuba Magnet Laboratory. http://www.nims.go.jp/TML/english/ Accessed June 17, 2011
5. Science in High Magnetic Fields. EuroMagNET II: research infrastructures for high magnetic field in Europe. http://www.euromagnet2.eu. Accessed June 17, 2011
6. Faculty of Science. High Magnetic Field Laboratory. http://www.ru.nl/hfml. Accessed June 17, 2011
7. Helmholtz-Zentrum, Dresden-Rossendorf. http://www.hzdr.de/db/Cms?pNid_580. Accessed June 17, 2011RIALS

Words and writing:

THE EVOLUTION OF THE PAGE

Until recently, communication was generally restricted to the capacity of the medium used to transmit it. About 5000 years ago, the Sumerians, Babylonians, and Assyrians used tablets for the purpose of written communication. These tablets were probably made of 2 materials: clay and stone (mostly limestone).[1] Clay tablets were portable, about 16_14 cm, could be hand-held, and sometimes were double-sided.

They were mainly used by students who indented the clay with Wedge-like instruments made from bone or wood. It is remarkable that the Sumerians did not break their texts to continue them on the opposite side of the tablet—that is, their intended text had to fit perfectly onto one side of a tablet. Clay tablets were then sun-baked; this process made them fragile and probably explained their relative scarcity. A benefit of sun drying was that the material could be recycled by soaking in water.

Communications intended for public display and those that should last longer were presented in a larger medium: the stone stele. The legal code of Hammurabi was written on a diorite stone standing over 2 m high. Steles were universal forms of public communications; fast forward nearly 4000 years and one finds them in the Mayan and other Mesoamerican cultures, fulfilling the same goal. Because it is relatively easy to carve, limestone was the preferred stele material in the ancient and new worlds. Romans improved on the tablet by making it out of wax. These were mainly used to teach writing to children, and their obvious benefit was that they could be easily "erased."

Writing was revolutionized by the creation of the scroll, a medium with larger capacity. Scrolls were built from papyrus, parchment, vellum, or paper. Individual pages of any of these materials were consecutively assembled to create a roll that could be meters long.[2] Scrolls were read horizontally from left to right, vice-versa, or both ways. Held together by wood rollers, one side was unrolled while the opposite was rolled to keep the reading area down to a comfortable size. These were the most popular media to store knowledge from antiquity (Alexandria and Rome) to the European Middle Ages. Similar to clay tablets, scrolls could be used on both sides (called an "opistograph"). Scrolls, such as the Torah, are still used in religious ceremonies. Jack Kerouac wrote *On the Road* on a scroll of paper. This famous modern scroll, now the property of the owner of the Indianapolis Colts (an American football team) has been displayed in museums around the world.[3] Hanging or vertical scrolls were used for ceremonial communications and are the direct predecessors of computer scroll reading and movie credits. Today, hanging scrolls are mostly decorative.

The next step in evolution was the accordion (or concertina) book. Accordion books probably originated in the Orient, Japan and/or China. They tended to be made of paper and offered a benefit over tablets and rolls in that they stood up by themselves, freeing the hands of readers (similar to your computer screen). In China, some accordion books were made of strips of bamboo. Nowadays the making of accordion books has been relegated to children, and some artists who use them as their medium to produce beautiful works of art.

Because papyrus is delicate, lasting only a few decades, and difficult to produce (the plant is found mainly along the Nile river), a different medium (parchment) was needed. The king of Pergamon is credited with having invented it (thus its name in Spanish: "pergamino"). Parchment is also somewhat difficult to make as it comes from the skins of animals. Although the Chinese invented paper way before the Common Era (CE), they did not use it as a writing medium until about 300 CE.[4] Around this time (200–400 CE), the first codices appeared in Central Europe. Codices are very similar to our current books in the sense that their sheets are attached on 1 side, allowing easy access to any portion of the book. Codices were initially made of parchment, but around 1200 CE, the Arabs brought paper to Europe where it was rapidly and widely adopted. Bookmaking was a personal activity, mainly done by monks, and thus literature was not widely accessible during the medieval times. Of course all of this changed in the mid-15th century when Johannes Gutenberg invented the printing press and bookmaking became an industry.

At this point in this short narrative, booklovers would fault me if I did not comment on incunables. "Incunables" are books manufactured before 1500 CE. They were printed from wood blocks or by means of the first movable type printing presses invented by Gutenberg. The word "incunable" comes from the Latin term for "cradle," implying the earliest of something. From a grammatic standpoint, it is important to note that words or text were not broken and continued elsewhere on a routine basis until codices and books were used.

A page is 1 side of a leaf or sheet of paper and, I think, a thing of beauty. When a book lies flat, the left page is called the "verso" and the right one the "recto" (verso meaning the reversed or back side of the leaf). Both together are called the "spread" of a book.

Recto pages have odd numbers, while verso pages are even-numbered (at 1 time in the past, only folios [see below] were numbered, not pages). Most book pages contain a "header" (also called a "running head") that generally refers to the title of the book or chapter and is displayed over the main text of each page.

"Footer" refers to material found at the bottom of a page, separate from the main text (some of the funniest footers I have read are found in a book of essays called *Consider the Lobster* by David Foster Wallace). *The size of books is historical in nature and originated about 300 years ago, because machines used then could cut or "trim" paper to only a certain size. If you take a piece of paper measuring 1 square meter and fold it in half 4 times as shown on the next page, you end up with a piece of paper (labeled L) measuring 210 _ 280 mm (8.5 _ 11 inches), which is the standard letter size and also a page size used for hardcover books produced in the United States (Fig 1).

If a letter-size sheet of paper is again folded in half (labeled T on the next page), one ends up with a paper of the size (135_216 mm or 5.3_8.5 inches) used for so-called trade paperback books (I am not sure why mass paperback book pages measure 110 _ 178 or 130 _ 198 mm). These differences in book sizes were apparently created to distinguish among them as they are sold in different fashions and at different prices. Larger books are thought to be more "literary," whereas the smaller ones are considered "good" values and are sold mostly in airports, train stations, supermarkets, and so forth.[5] The size of "pocket" book pages is sometimes capricious, and as an example, I offer the French publisher Actes Sud. Its founder, Hubert Nyssen, measured his hand from the bottom of his palm to the tip of his index finger and from the outer aspect of the hypothenar eminence to the outer aspect of his thenar region. If he could hold this size paper in his hand, it must be perfect for all! The folding method illustrated above also establishes the size of common sheets of paper used in the United States (from smaller to larger: letter, legal, junior legal, ledger, tabloid, and so forth). These sizes are slightly larger than those found in books, and I will explain why below.

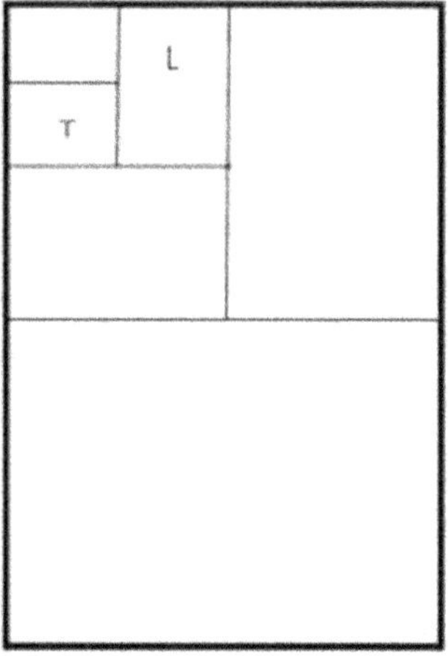

Fig 1. Diagram illustrating the normal dimensions for standard letter size (L) and trade paperback books (T).

Books (and the *American Journal of Neuroradiology* [*AJNR*]) is considered one strictly speaking) are composed of concertina-like "signatures." The name given to each signature depends on its length: folio equals 2 pages; quarto, 4; sexto, 6; octavo, 8, and so forth (confusion arises because the same terms are sometimes used to describe different paper sizes). The largest signatures are 32 pages because this is the traditional limit for machines that stitch or glue them together. If you have ever seen a signature, you know it is weird and at first glance makes no sense. Folio signatures look like this[6]:

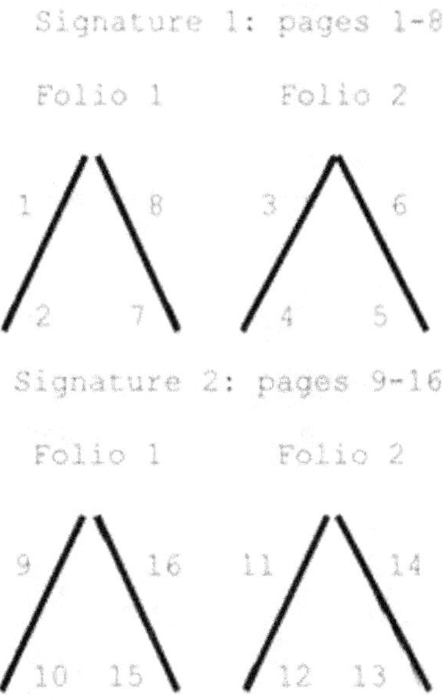

For signature 1, folio 2 will be placed into folio 1, creating continuous pages once the outer edges (folds in larger signatures) of the pages are cut or trimmed. Trimming involves removing about one-eighth of an inch around the sides, thus bringing down the measurements previously given for paper (vide supra) to what you find in books. The left side of the signatures is bounded by stitching or gluing. Stitching can be done by over sewing or sewing through the fold (the preferred method so that books can open flat). Conversely, glue is applied to the angle of the signatures, and then these are bound to a piece of cloth or paper (as is done for *AJNR*). The problem with this type of binding is that with time the signatures may become loose and separate. In any book, journal, or magazine, the cover has 4 consecutive parts: front or first, front inner or second, back inner or third, and back or fourth cover. Hardcover books (at least in the United States) often have dust jackets that also have front and back covers as well as front and back flaps. In journals and magazines, the second, third, and fourth parts of the cover are premium advertising space (in books dust jackets provide extra space for reviews, summaries, and biographies). A flyleaf is that empty leaf at the start and end of a book (not found in journals or magazines). Regular pages have a central print space surrounded by blank margins.

The framework of a page is not a fickle choice. Methods used to determine the most eye-pleasing placement of text include those of van der Graaf and Tschichold (see below, red square is the print space, illustrated for only 1 side of the spread) (Fig 2).

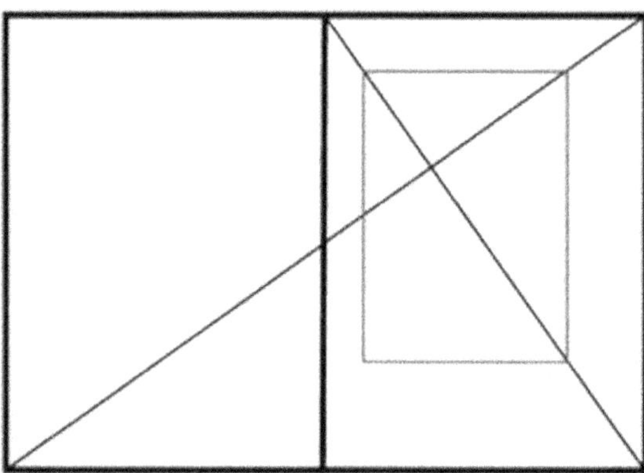

Fig 2. Diagram illustrating method used to determine the most pleasant dimension of text margins.

The above diagram illustrates the major rules of thumb regarding page margins: different sizes for all 4. The outer and bottom ones are double the size of the inner and top ones, respectively. This is also known as the "golden ratio" for page margins and creates the most eye-pleasing ones. Journals and magazines where space is at a premium sacrifice aesthetics and nearly always have uniform margins with the text close to the edges of the paper (Jack Kerouac left no margins in his scroll).

Some pages contain columns of text (generally 2, but up to 4 are common and 3 are said to be the most pleasant layout).

This is done to break down long lines of text (rule of thumb: a line longer than 60 characters is difficult to read).[7] How wide should a column be? Another rule of thumb is that the point size of the font multiplied by 2 equals its width in picas (1 pica is about 4.3 mm). Thus, the *AJNR* font size is 9.5, times 2 equals 19, times 4.3 equals 82 mm, which is the width of our columns (go ahead and try it).

"Marginalia" refers to notes, comments, and other writings on page margins. Notations by famous persons can add value to a book while those that you or I create probably will not.

Computer programs such as Microsoft Word allow comments on the margins of pages and some even allow Web-sharing of this type of marginalia (called "shared annotations"). E-readers such as Amazon's Kindle are capable of allowing marginalia and other annotations and go further by permitting you to share them with other readers via social networking programs.

E-reader pages reproduce basically all aspects of paper, albeit at a higher cost and with increased difficulties when it comes down to marginalia (it is easier to scribble on the margin of a paper page than to type there by using the tiny physical or virtual keyboard found in E-readers).

E-readers try to give readers the same experience as print. (I often ask myself why. If I buy an E-reader, it is to have a completely different experience; if not, why not buy the book?) Of course, with digital books, space is no longer a consideration as it was with print and most have variable page and font sizes.

Adobe PDFs also imitate the print experience as their software allows you to "leaf" though a document that simulates printed pages. Leafing is a feature found on the iPad (Apple, Cupertino, California) (something that I find anachronous and bothersome while I have heard others say it is cute). Your computer screen is basically a vertical never-ending virtual scroll in which the length of the text matters little. Paradoxically, as the capacity to store and display the written word increases, our attention spans decrease and make it impossible to take full advantage of these media.

The evolution of the page seems to have closed the circle: from clay tablets to computer tablets (witness the iPad), from ancient scrolls to scroll computer reading, from parchment pages to E-paper and E-readers. The capacity of the page no longer determines its contents. Margins are becoming superfluous as small screens must be completely filled with text to take maximum advantage of their viewing areas. Marginalia is no longer private with E-readers. I am not a fanatic of a specific medium, but as long as you are reading something, it will be contained within pages.

* N.B. This is what a footer looks like. It is not commonly used in scientific writing and perhaps it should be. N.B. stands for "nota bene," which instructs the reader to pay attention to something. Endnotes are not the same thing as footers or footnotes. Endnotes appear separately at the end of an article, essay, chapter, or book. If you need to call attention again to something in a footnote, you can add another N.B. as follows: N.B.2

References

1. Earth's Ancient History: A Website Dedicated to Ancient Times. http://www.earth-history.com/Sumer/Clay-tablets.htm. Accessed December 22, 2010
2. Scroll. Wikipedia. http://en.wikipedia.org/wiki/Scroll. Accessed December 22, 2010

3. Boston.com. http://www.boston.com/news/local/articles/2006/07/27/kerouacs_ road_will_be_unrolled. Accessed on December 22, 2010
4. History-of-Books. Wikipedia. http://en.wikipedia.org/wiki/History_of_books. Accessed December 22, 2010
5. guardian.co.UK. http://www.guardian.co.uk/books/2001/aug/11/gettingpublished. Accessed December 22, 2010
6. http://icpcres.ecs.baylor.edu/onlinejudge/external/9/999.html. Accessed December 22, 2010
7. Column (typography). Wikipedia. https://en.wikipedia.org/wiki/Column_(typography) Accessed December 22, 2010

WRITER'S BLOCK

After writing nearly 70 *Perspectives*, it was bound to happen; it seems I have hit "writer's block." This is the first time I sat down to write my monthly essay in over 8 months and I decided to use the occasion to familiarize myself with this condition (and try to liberate myself from it). Writer's block varies in intensity from extreme (abandoning one's career as an author—think Harper Lee and Ralph Ellison) to trivial and temporary (which I hope mine is). Why worry? Most writers have it at one time or another.

The most common causes cited for writer's block are lack of inspiration, illness, depression, financial pressure, and a sense of failure. None of these apply to me. In my case, maybe it is just good old academic pressure. With increasing frequency, I hear that the only part of *AJNR* that our subscribers read is my *Perspectives*, to which I say: What about the rest of the *Journal*? That is where most of my energy and time are spent! And, if our readers are paying that much attention to my short essays, should each one be better than the last? Writer's block sounds like the ideal excuse to avoid thinking about what to write (just look at the title of my first reference: "Writer's Block as an Instrument for Remaining in Paradise").[1]

Writer's block is a modern notion, and the term was coined in 1947 by Dr Edmund Bergler, a famous Austrian psychiatrist living in New York City.[2] Today, it is well accepted that the notion of writer's block arose in conjunction with the sudden prestige of psychiatry in the United States after World War II. Dr Bergler, a follower of Freud, blamed writer's block on oral masochism and a milk-denying mother (that gives me something to think about because I know that I was bottle-fed!), in addition to other "phallic and anal" explanations along Freudian lines. Stress leads to panic, and some scientists believe that the reticular activating system in the brain stem will shift higher functions associated with writing from the cortex to the limbic system under duress.[3] Others disagree and think that the creative writing process starts at the level of the limbic system, whereas more technical writing is initially fueled by the frontal cortex. If both were true, all writing would stop as functions shift from one location to the other. However, writer's block can be selective, as it is in my case. That is, I continue to write other articles, chapters, and books, but writing this specific series of essays is my problem. Writer's block is commonly seen in college and university students who consistently fail to turn in their written assignments. In them, procrastination (a behavior specifically called "academic trait procrastination") is a major component of writer's block. Procrastination is learned, so education specialists claim that it can be unlearned.[1] Perfectionism is also blamed for the block; it seems to motivate some procrastination, and together these both promote writer's block.[1]

In academia, the notion of writer's block is disdained by younger members but seems to gain respect at higher levels where it occurs more commonly.[4]

Writer's block is better termed "creative inhibition" or "creative block."[5] It is becoming more prominent: it was little known by the early Romantic writers, became more prominent during the epoch of the French Symbolists, and last, was rampant (and became a recognized entity) during the period of the great American novel. Today, in a manner similar to attention-deficit disorder, writer's block is a nearly unique American affliction (though it occasionally happens in other countries, vide infra).[6]

Agraphia is the ultimate writer's block because it refers to the physiologic inability to write, but, in this case, lesions in the brain, such as those induced by trauma or stroke, are present and explain it. Agraphia results from damage to the Wernicke area and is nearly always accompanied by other language disabilities. In some cases, the inability to write may be physical, such as so-called "writer's cramp." This is a muscle dystonia, and DTI has shown fractional anisotropy changes in the fibers connecting the primary sensorimotor areas with subcortical structures in individuals who suffer from it.[7] In such individuals, fMRI has also shown abnormally low activation of the sensorimotor cortex and supplementary motor areas.[8] The findings of these studies imply that both inhibitory and excitatory mechanisms play a role in writer's cramp and that the pain it causes prevents writing by hand. Writer's cramp can also develop during typing and other activities such as using a screwdriver. Compared with men, women are thought to be better writers; however fMRI does not show significant differences in brain activation for either sex while writing.[9]

The same study found significant differences between good and poor writers while handwriting, mostly in brain regions involved with planning for serial finger movements.

The opposite of writer's block is also known to occur, and it can be temporary or affect an individual all of his or her life.

Balzac, Hugo, and Dickens probably had "graphorrhea." The problem with calling the obsession to write "graphorrhea" is that this term is also used for manic patients who compose long lists, many times containing only meaningless words, which is not the same as writing many great novels. As psychiatry evolved from an analytic discipline to a chemical science, writer's block came to be blamed on abnormal brain chemistry. More seriously, writer's block can be a manifestation of a dangerous underlying psychiatric disease such as schizophrenia, obsessive compulsive disorder, or substance abuse (think Scott Fitzgerald).

Writer's block has been assessed in individuals who speak languages other than English. Two studies addressed it in Chinese and Spanish native speakers and found that it occurs in those languages as it does in English.[10-12] In other languages, as in English, writer's block appears to be related to premature editing and to a lack of strategies for dealing with complex writing tasks. Developing a strategy before the actual writing helps some individuals.

Princeton Professor and author John McPhee tells a related story in his essay "Structure."[13] For 2 weeks he lay down on a picnic table under the trees looking at them and wondering how to start a piece on pines. After 8 months of work, he was finally able to turn it in. He suggests that having a preplanned structure eases the stress of writing and results in a better organized and flowing article. The ease of cutting and pasting makes attaining the desired structure easier today than in the past.

The use of a computer with word-processing capabilities reduces writer's block for second-language writers but not for native- language writers.[14] Spelling is also intimately related to the ability to write. In one study, disabled spellers showed significantly more activation in clusters of neural networks associated with working memory and executive functions.[15] Computer programs that automatically correct spelling may help these individuals overcome writer's block.

Because writing is intimately related to reading, recognition of the written word is needed for both activities. Alexia is a condition in which patients lose their ability to read and is associated with lesions in the left parietal and occipital lobes.[16] Alexia has been "folklorized" in several accounts by the famous and popular author and neurologist Oliver Sacks. His patients who had this condition were said to have lesions affecting the VWFA (visual word recognition area), which is presumably located in the left midfusiform gyrus (running from temporal to occipital lobe under the parahippocampal gyrus). Because this area is also involved in the recognition of auditory, phonologic, and visual impulses, patients with lesions there have more symptoms than alexia only. Pure alexia caused by a lesion in the VWFA has not been reported.

More importantly, a meta-analysis of the literature, including fMRI studies, states that this brain region does not participate in visual word processing, and thus its concept is erroneous.[17] Alexia without agraphia occurs with lesions involving the left-sided splenium of the corpus callosum.

Strategies for coping with (and curing) writer's block include group discussions, brainstorming (I wrote an essay against this), list-making (I have a long list of topics that I have considered, but none seem very attractive now), and engaging with the text (I have no idea what this means). Recalcitrant blockage must be treated with extensive "therapy."[15] Other cures include "automatic writing" in which texts are produced from the subconscious without conscious awareness.[18] Instead of coming directly from the writer's mind, Arthur Conan Doyle believed that automatic writing came from external spirits. Channeling writing from a spirit is called "psychography." Both phenomena may be explained as "ideomotor effects" meaning just an activity of which we are partially or completely unaware. Of course, all of this is nonsense, and fMRI has proved (many times) that ideomotor effects originate in the brain and not outside of it.[19]

At this point, I must say that having finished this *Perspectives*, I feel somewhat liberated. Whether that feeling will be short-lived or last and allow me to write next month's piece, you, kind reader, must wait and see. k

REFERENCES

1. Smeets S. **Writer's block as an instrument for remaining in paradise.** *Zeitschrift Schreiben* 2008 Jun 22 [Epub]
2. http://en.wikipedia.org/wiki/Edmund_Bergler#cite_note-Akhtar2009–11. Accessed July 24, 2013
3. Bane R. **The writer's brain: what neurology tells us about teaching creative writing.** http://www.rosannebane.com/uploads/2/5/6/2/25628316/the_writers_brain.rosanne_bane.41-501.pdf
4. Gumz A, Brahler E, Erices R. **Burnout experience and work disruptions among clients seeking counseling: an investigation of different groups of academic level.** *Psychother Psychosom Med Psychol* 2012;62:33–39
5. Kantor M. *Understanding Writer's Block: A Therapist's Guide to Diagnosis and Treatment.* Westport, Connecticut: Praeger; 1995
6. Acocella J. **Blocked: why do writers stop writing?** *The New Yorker* June 16, 2004
7. Delmaire C, Vidailhet M, Wassermann D, et al. **Diffusion abnormalities in the primary sensorimotor pathways in writer's cramp.** *Arch Neurol* 2009;66:502–08
8. Oga T, Honda M, Toma K, et al. **Abnormal cortical mechanisms of voluntary muscle relaxation in patients with writer's cramp: an fMRI study.** *Brain* 2002;125:895–903

9. Richards TL, Berninger VW, Stock P, et al. **fMRI sequential-finger movement activation differentiating good and poor writers.** *J Clin Exp Neuropsychol* 2009;31:967–83
10. Lee S, Krashen S. **Writer's block in a Chinese sample.** *Percept Mot Skills* 2003;97:537–42
11. Betancourt F, Phinney M. **Sources of writing block in bilingual writers.** *Written Communications* 1990;7:482–511
12. Phinney M. **Word processing and writing apprehension in first and second language writers.** *Computers Composition* 1991;9:65–82
13. McPhee J. **Structure: beyond the picnic-table crisis.** *The New Yorker* January 14, 2013, pp 46–55
14. Richards TL, Berninger V, Winn W, et al. **Differences in fMRI activation between children with and without spelling disability on 2-back/0-back working memory contrast.** *J Writing Res* 2009; 1:93–123
15. Huston P. **Resolving writer's block.** *Can Fam Physician* 1998; 44:92–97
16. Alexia (condition). http://en.wikipedia.org/wiki/Alexia_(condition). Accessed July 24, 2013
17. Price CJ, Devlin JT. **The myth of the visual word form area**. *Neuro-Image* 2003;19:473–81
18. Automatic writing. http://en.wikipedia.org/wiki/Automatic_writing. Accessed July 24, 2013
19. Spengler S, von Cramon DY, Brass M. **Was it me or was it you? How the sense of agency originates from the ideomotor learning revealed by fMRI.** *NeuroImage* 2009;46:290–98

QWERTY, @, &,

Editing is a mostly lonely activity, with many hours spent staring at the computer screen and keyboard. Although most editors I know are men, women are much more facile with words, spoken and written. Women speak an average of 7000 words per day, whereas men average only about 2000.[1] This difference is even present in young girls, who tend to be more linguistically gifted than their male counterparts.

Additionally, women are better communicators because they have a larger catalog of facial expressions and body movements.

Regardless of sex, if we type the way we talk, all of us do a lot of typing. For most of us in the medical field, typing is an activity as important as speaking and one of the most practical and useful skills to possess (I have come to the conclusion that I learned only 2 truly practical things while in elementary and high school: English and typing). The average person can easily type between 30 and 70 words per minute, and, as they say, practice makes perfect; or as the famous nonfiction author and Princeton professor John McPhee said, "Writing teaches writing."[2]

To write, most individuals nowadays use a keyboard (very few still handwrite). The computer keyboard became popular in the early 1980s and was adapted from IBM's widely used electric typewriters. Initially, it had 83 keys of varying sizes, closely spaced without clear horizontal or vertical separations.[3] In this first computer keyboard version, there was no alphanumeric keypad and the function keys were grouped on the left side. The next iteration, called the 84-key version (also known as the AT keyboard), added the alphanumeric pad on the right side. In the mid-1980s, the number of keys was expanded to 101 and the function keys were relocated to the top row where they remain. Variations with 102 or 103 keys were briefly manufactured before settling on the current 104 keys (the extra ones generally serve to access specific Windows functions). So-called ergonomic keyboards split the keys at the middle, curving and separating them slightly, something I cannot get used to. The keyboards in notebook or portable computers have been redesigned for economy of space, making them somewhat more difficult to use. Another problem is that there are no standardized arrangements for keyboards on portable computers, making them all slightly different.

The most common keyboard layout, used for the English language, is called "QWERTY." This name comes from the order of the first 6 keys in the upper left-hand corner of the keyboard. Why the keys are arranged this way no one is sure. Around 1875, the first keyboards for typewriters were created and their keys were alphabetically arranged. It is said that Christopher Latham Shore, an early inventor of the typewriter, changed them to the QWERTY configuration to make it more difficult to type fast and jam the key bars.[4] Because we are all used to this somewhat bizarre key arrangement, manufacturers do not want to change it.

In the mid-1930s, Dr. August Dvorak, an efficiency expert, studied the way people type and created a different keyboard arrangement (called the American or Dvorak Simplified Keyboard) that groups the most commonly used keys in the areas that are the easiest to reach. Experts prefer it because it is more efficient: In the regular keyboard, about 30% of key presses occur centrally against 70% in the Dvorak type. Muscle fatigue from typing is thus minimized by using the Dvorak keyboard. The truth is that, unfortunately, it is very difficult to switch to the Dvorak keyboard after learning the QWERTY one. Many argue that because standard computer keyboards are direct descendents from old typewriters, they are not matched to the potential offered by computers. The most typical examples of this inheritance are the wide space bar located in the lowermost row of keys and the shift key. In typewriters, the latter was used to shift the carriage (or basket) so that the part of the type bars containing capital letters would strike the paper (thus "shift lock" is now more appropriately called "caps lock"). Another change needed in the computer era was the addition of separate keys for zero and the number 1 (before, you could use the letter *O* and a capital *I* for zero and 1, respectively).

The fact that keyboards used to be attached to computers via wires never bothered me, but companies decided to give their customers greater freedom (to type where?), and most keyboards nowadays use radio-frequency (including Bluetooth) or infrared devices to achieve this autonomy.

Not only are most inexpensive wireless keyboards slower than their wired counterparts, you also need to buy batteries for them and for your wireless mouse. Of course, as voice recognition improves, keyboards may begin to lose their importance.

Most European countries use the same standard alphabet and Roman numerals used in English (called American Standard Code for Information Exchange or ASCII characters).

In the German keyboard, the only difference is that *Y* and *Z* positions are swapped, leading to the QWERTZ keyboard arrangement. For French-speaking countries, the keys for *A*, *Q*, *Z*, *W*, M, and *N* are in different locations (hence the AZERTY keyboard). Because languages other than English commonly use accents, a combination of key strokes is needed to display them using the QWERTY keyboard.

ASCII includes 31 characters. These are the characters found in most keyboards (full-sized or portable) throughout the world. Other characters no longer used or adapted to the ASCII code are included in the Extended Binary Coded Decimal Interchange character set (designed by IBM) and the z/OS UNIX.[5] Because different alphabets have different character counts, keyboards in other languages have a variable number of keys (101–105). A reduced number of keys is found in mini-keyboards, sometimes called "thumb boards" or GKOS keyboards (such as the ones for BlackBerrys and similar devices). "Dead keys," found in some keyboards, may be "revived" by assigning them specific functions such as accents or special characters.

The additional alphanumeric (or simply numeric) pad of the keyboard is found only on desktop computers and, for purposes of space-saving, is not included in most portable ones. Most conventional keyboard functions are accomplished by mechanical levers and electronic switches.

Other variations of keyboards include those found in touch screens, which are becoming more popular with tablet computing. Touch screen keyboards are considered the natural evolution of "on-screen" keyboards, in which an image of the keyboard appears on the screen and keys are selected by clicking the mouse. Foldable or flexible keyboards are made of plastic or silicone and are great for traveling. They can be attached to computers and other devices such as cellular telephones.

Flexible keyboards are also ideal for hospitals and laboratories because they can be washed and disinfected, and the absence of crevices between keys makes them "cleaner."

Britain's *Daily Mail* newspaper reported that computer keyboards have more than 150 times the acceptable number of germs and are 5 times dirtier than a toilet (a fact to keep in mind when you are eating your sandwich while typing or surfing the Web). In a study performed here at the University of North Carolina, computer keyboards housed in the Burn Unit were found to be uniformly infected with coagulase- negative *Staphylococcus* organisms, a common source of hospital-acquired sepsis.[6] Diphtheroids were present on 80% of those keyboards, and are particularly dangerous for immunosuppressed individuals such as those with extensive burns. Commercial cleaners maintain keyboards bacteria- free for about 48 hours. A benefit of one of the most intriguing new keyboards, the holographic or projection keyboard, is that the flat surface used for its projection can be easily cleaned. A laser projects an image of a keyboard onto any flat surface, detects keystrokes, and even simulates the clicking noise of a conventional keyboard. These are truly virtual keyboards, and miniature versions that can be used with smart phones have just hit the market. It does not matter which keyboard you use or prefer as all contain some bewildering keys.

One of the most commonly used keys is the "at" symbol, @, which shares the number 2 key in the QWERTY arrangement.

@ means simply "at," "located at," or "at the rate of." @ has been present in keyboards since 1885 but became ubiquitous in the early 1970s when used in the first e-mail messages. In other languages, the @ symbol is more colorfully named (eg, "snail" in Italian, "monkey tail" in German, "dog" in Russian, and "little mouse" in Chinese).[7] In Spanish, Portuguese, and French, @ denotes an old measure of weight (the arroba) and is called "arrobas" or "arrobase" (French). @ is probably of Italian origin and was initially used by Venetians to designate the amount of weight contained in an amphora. Currently, @ is most commonly used in e-mail addresses to separate the name of a person from the domain in which the address is located. In text messaging,@may serve as a substitute for "at."

Recognizing the importance of @, in 2010 the Museum of Modern Art in New York City admitted this sign into its architectural and design collections.[8]

Although substituting @ for "at" does not save me many keystrokes, using "&" instead of "and" is more economical.

The ampersand, &, means "and per se and" or more simply "and." & dates back to the first century of the Common Era and its shape has been progressively changed by the Romans and French. The ampersand should not be used to mean "et," which is generally symbolized by "7." When handwritten, the ampersand looks a bit different: (sometimes the vertical line is left out). Regardless of its exact shape, I think the ampersand is one of the most elegant and practical symbols used in language.

The number sign, #, is probably used as commonly as @ and &. It is usually used to designate a numeric position such as the following: *AJNR* is the #1 journal in clinical neuroimaging.

In the United States, # is called the "pound" sign, whereas in other countries, it is simply known as the "number" sign (scientists sometimes call it the "octothorpe").[9] Calling it a "pound" sign may lead to confusion in England, where the pound sign is £. Thus in England, # is called the "hash" sign. In Spanish-speaking regions, the number sign is generally "No."

In Spanish, # has many names ("almohadilla," "cardinal," and even "tic-tac-toe"). The musical symbol "sharp" is nearly identical to #, but its 2 horizontal bars are angled upwards from left to right. A fact that is interesting to editors is that in copyediting, ### means that more content will be added or that mistakes that need to be corrected are found in the text. ### at the end of a manuscript means no further information is forthcoming. Chess fans know that # after a move means "checkmate."

Last, a few words about keyboards and health. I now spend more hours in front of my computer screen typing than ever before. Strain to your wrists, arms, back, and neck from typing may cause pain. Keep your shoulders in a relaxed position, your elbows at about a 90° flexion, and your wrists and back straight. Get to know your keyboard and play it like a piano:

Do not rest your palms or wrists on anything. Take short and repetitive breaks throughout the day. They are good for the body and the mind.

References

1. http://itre.cis.upenn.edu/_myl/languagelog/archives/003420.html. Accessed June 3, 2010
2. McPhee J. **The art of nonfiction No. 3**. *The Paris Review* Spring 2010:192
3. **Standard keyboard layouts.** The PC Guide. http://www.pcguide.com/ref/kb/ layout/std.htm. Accessed June 3, 2010
4. **QWERTY alphanumeric layout.** The PC Guide. http://www.pcguide.com/ref/ kb/layout/alpha_QWERTY.htm. Accessed June 3, 2010
5. IBM WebSphere Application Server. http://publib.boulder.ibm.com/infocenter/zos/basics/index.jsp?topic_/com.ibm.zos.zappldev/zappldev_14.htm. Accessed June 3, 2010
6. Rutala WA, White MS, Gergen MF, et al. **Bacterial contamination of keyboards: efficacy and functional impact of disinfectants.** *Infect Control Hosp Epidemiol* 2006;27:372–77. Epub 2006 Mar 29
7. **At sign.** Wikipedia. http://en.wikipedia.org/wiki/At_sign. Accessed June 3, 2010
8. **Why @ is held in such high design esteem.** *The New York Times.* http://www. nytimes.com/2010/03/22/arts/design/22iht-design22.html?ref_technology. Accessed June 3, 2010
9. **Number sign.** Wikipedia. https://en.wikipedia.org/wiki/Number_sign. Accessed_September_30,_2015

PREDATORS AND CRANKS

A few years ago I received an e-mail invitation to write a review article on MR spectroscopy from a journal I did not know.

Thinking that it was a good project for one of my visiting research fellows, I accepted. It took us about 3 months to write and illustrate the article, and when we submitted it, we signed, as is commonly done, their copyright agreement. Much to my surprise, we shortly thereafter received an invoice for US $2700 because this journal operated under the "open access" model. Paying that much would have used much of my "book and travel" allowance, so we retracted the publication only to find that retractions were impossible! After much back and forth and threatening to get the University's lawyers involved, they returned the article to us (it was later published in the *Neuroimaging Clinics*, a bona fide journal, not open access). While there is nothing wrong with open access and the *American Journal of Neuroradiology* (*AJNR*) supports it as long as our current financial model remains stable, it is a system ripe for abuse by many.

In April 2013, the *New York Times* (*NYT*) published a piece on "predatory" medical journals and scientific meetings.[1] It described how a group of scientists were duped into participating at a meeting that initially seemed legitimate (see below). Welcome to the world of "pseudoacademia," where newly created outfits recruit speakers and authors strictly for profit in activities that are not linked to any respectable scientific society, group, or journal.

There are currently hundreds of companies that "sponsor" meetings and journals under the rubric of "open access." The open access movement arose from the need to share information with all of those who are interested in it while trying to avoid paying for subscriptions or buying individual articles, especially if these sprang from investigations financed by public funds. Although controversial, open access makes articles easier to find and quote and is thus beneficial to authors and readers alike. There are legitimate, prestigious open access journals such as ones published by the *Public Library of Science*, which rightly demand a fee for publication. Predatory publishers take advantage of the open access movement and of our never-ending hunger to fill our resumes and be promoted and have developed journals that closely resemble genuine ones. In September these fraudulent activities hit home when many of us received an e-mail message from Ivy Union

Publishing Company (a well-known publisher of predatory journals) recruiting editorial board members for the new *AJNR* (*American Journal of Neuroscience Research*). The e-mail was designed in blue tones similar to those used by our *AJNR*, and even the font used was exactly the same as the one in our previous cover design. Immediately we contacted our lawyers, who sent a letter to Ivy Union Publishing demanding that they cease to use our trademark, to which, not surprisingly, we have yet to receive a response.

Jeffrey Beall, an academic librarian at the University of Colorado, has created a list of predatory journals (commonly known as "Beall's list") and their publishers.[2] Mr. Beall divides these publications into those questionable publishers that have portfolios of up to hundreds of journals (the Ivy Union Publishing Web site had 131 pages of journal titles when I looked at it) and individual journals published outside traditional platforms. Before submitting an article to a new journal, Mr. Beall suggests that one checks his list of criteria for determining predatory open access journals and publishers found at http://scholarlyoa.com/2012/11/30/ criteria-for-determining-predatory-open-access-publishers-2ndedition.

Briefly, any of the following should steer one away from submitting articles to a journal:

1. The name of the journal is incompatible with its scope.
2. Its national base is not clear.
3. Submission-to-publication periods are incompatible with traditional peer review (my comment: less than 21 days is suspicious).
4. No clear editor and no editorial board.
5. No Impact Factor listed.
6. Unprofessional, hastily put together Web site.
7. No mention of fees until an article has been accepted.

Today, there are more than 4000 predatory journals that publish 10%–15% of all open access articles. Not only are they tricking authors into submitting and paying for their articles, they offer members of their editorial boards as much as 20% of the author fees. Once you become a member of one of these editorial boards, it is basically impossible to be removed as several anecdotes in the *NYT* article recount.[1] Needless to say, Mr. Beall is being sued by several of these publishers, has been a victim of vicious on-line comments, and is the subject of Internet campaigns to discredit him.[3]

Because rapidly developing economies are generating a significant number of new researchers, most "open access" publishers are springing up there, but their Web sites manipulate the truth to appear as if they are headquartered in the United States, United Kingdom, Australia, or Canada (though no contact information is found on the Ivy Union Publishing Web site, our investigation led to an address in Delaware, which then led us to an address in Boston). The geographic bases of predatory journals can be found on Semantico.com. Data there show that by plugging the IP addresses of 192 predatory journals and 321 predatory publishers into a geolocator, one finds that 65% of such journals and 67% of publishers were registered in the United States. Because one can never be sure whether the locations linked to the addresses are real or fake, it is always possible that indeed these publications actually started in the United States.

Predatory journals care little about the quality of science and are known to sometimes publish plagiarized work.[4] Their articles receive little professional formatting to save costs, and they are never listed in the larger citation databases such as PubMed, Web of Science, or Scopus; a fact that nullifies their open access spirit because they are very hard to find and quote. Sometimes, predatory journals even publish articles without the author's permis- sion. Many of these journals offer to translate their articles into 50 different languages, when, in reality, this feature only directs the readers to use the free Google Translate service for this purpose.

A glaring example of types of peer review and acceptances carried out by predatory journals is illustrated by the following hoax. Two well-regarded professionals utilized SCIgen (http://pdos.csail.mit.edu/scigen), a computer program that generates nonsensical articles dealing with computer science, to create an article that was submitted to a predatory journal called *The Open Information Science Journal*.[5] The authors even gave their affiliation as the Center for Research in Applied Phrenology (CRAP)!

After the article was accepted, they received a bill for US $800 to be sent to a POBox located in the United Arab Emirates. However, to be fair, SCIgen was also used to generate a similar article that was submitted and accepted by a reputable journal published by Elsevier.

The problem gets worse, and the lines, blurrier. Between 2000 and 2005, publishing giant Elsevier published 6 fake medical journals, all sponsored by pharmaceutical companies, and as if that was not bad enough, these journals often contained reprinted articles that were favorable to products manufactured by the sponsoring companies.[6] Immediately after this was publicly disclosed, the CEO of Elsevier's Health Sciences Division issued an apology and a reassurance that this would not occur any more.

All of the above also extends to congresses and meetings. In the previously mentioned *NYT* article,[1] researchers were tricked into presenting at a meeting called "Entomology-2013" when they thought they were presenting at the well-recognized and prestigious "Entomology 2013" (do you see the difference in the titles?).

Later they were charged for participating at the meeting. Last month I received an invitation to participate in the 1st International Conference of Radiology to be held in Raleigh, North Carolina (just 30 miles away from where I live). The invitation that came from some outfit located in China promised me time at the podium, dinners, "mingling" with the best researchers, and a name badge that would clearly identify me as a prominent participant and world expert. When I did not respond, I was bombarded with spam-like e-mail messages asking me to confirm my participation.

These so-called "crank" meetings promise luminary speakers who often do not have enough valid publications to support this denomination or simply have not published their research.[7] To me, it is not clear who attends and who lectures; I do not know anyone who has.

One of the best known crank meetings is Autism One (which is held in Canada and the United States).[7] In it, researchers of dubious integrity give talks, and the main speaker is generally Jenny McCarthy. Ms. McCarthy, a former *Playboy* Playmate, is a popular television show host and author of books on parenting, alternative medicine, and autism. Other guest speakers generally have published their results in blogs and popular media and, at best, in predatory journals. Unfortunately, serious institutions such as the University of Toronto, the Sick Kids Foundation, and even the American Academy of Pediatrics have been suckered into debacles stirred by presentations at these autism meetings.

I urge you to look in your own backyard for predators. Many institutional libraries, when choosing journal subscriptions, have rules that force them to buy those under the categories of "Gold or Green Open Access" (one archives the articles for the authors; in the other, the authors themselves archive the articles) or those that adhere to the Creative Commons license agreement. Gold or green has nothing to do with paying to get published, just with access. Most predatory journals claim to be gold or green to make themselves attractive to libraries. Like true predators, these journals stay around just while there is prey. Anecdotally, I heard that a predatory publisher abruptly closed its doors (and Web site) once its profits reached US $100 million. Beware, because predatory journals will take away not only your money but, more important, your prestige, reputation, and self-respect.

REFERENCES

1. Kolata G. **Scientific articles accepted (personal checks, too).** *New York Times*. April 7, 2013. http://www.nytimes.com/2013/04/08/ health/for-scientists-an-exploding-world-of-pseudo-academia. html?pagewanted_all&_r_0. Accessed September 16, 2013
2. Beall's list of predatory, open-access publishers. http://www.academia.edu/1151857/Bealls_List_of_Predatory_Open-Access_Publishers. Accessed September 16, 2013
3. Butler D. **Investigating journals: the dark side of publishing.** *Nature* 2013;445. http://www.nature.com/news/investigating-journals-thedark-side-of-publishing-1.12666. Accessed September 16, 2013
4. Sanchez J. **Predatory publishers are corrupting open access.** *Nature* 2012;489. http://www.nature.com/news/predatory-publishers-arecorrupting- open-access-1.11385. Accessed September 16, 2013
5. Aldhous P. **CRAP paper accepted by journal.** *New Scientist* 2009. http://www.newscientist.com/article/dn17288-crap-paper-acceptedby-journal.html#.Ujc7HD-Jnq5. Accessed September 16, 2013
6. Grant B. **Elsevier published 6 fake journals.** *The Scientist* 2009. http://www.the-scientist.com/?articles.view/articleNo/27383/title/Elsevier-published-6-fake-journals. Accessed September 16, 2013
7. Gorski D. **Crank "scientific" conferences: a parody of science-based medicine that can deceive even reputable scientists and institutions**.

Science-Based Medicine 2009. http://www.sciencebased medicine.org/crank-conferences-a-parody-of-science-basedmedicine- that-can-suck-in-even-reputable-scientists-and-institutions. Accessed September 16, 2013

THE FRAUD AND RETRACTION EPIDEMIC

A recent article in the *Proceedings of the National Academy of Science (PNAS)* examines the cause of retractions involving more than 2000 articles published in biomedical and life-science related journals.[1] Of these, nearly 70% were retracted due to author misconduct, with the most common problem being suspected fraud (43.4%), followed by duplications and plagiarism.

When compared with data obtained in 1975, the incidence of misconduct-related retractions has increased 10-fold.

Overall, misconduct-related retractions involve only a tiny portion of the more than 25 million articles housed on PubMed.

The issue is not that the number of retractions is small but that their number is increasing considerably and rapidly. Exactly how many articles are retracted due to misconduct is difficult to establish as published retraction notices are often vague and unclear as to the cause of the problem (estimates place the figure at 0.2% of 1.4 million articles annually published). Whereas from 2002 to 2006 fraud-related article retractions were 20% higher than error elated ones, from 2007 to 2011 error-related retractions were less than 40% of fraud-related ones.[1]

It is hard not to point fingers; most fraud-related retractions come from our own backyard: the United States (probably reflecting the fact that about 26% of all scientific publications originate here). China and India account for the most retractions due to duplications and plagiarism (probably reflecting difficulties with the use of English). China's share of scientific publications went up from 4.4% to 10.2% from 2003 to 2008, while those from the

United States and United Kingdom went down, positioning China to become the largest source of origin in science in the near future.

Another interesting observation made by Fang et al[1] refers to the quality of journals in which most retractions happen. There was a direct correlation between Impact Factor (IF) and retraction numbers. Prestigious journals such as *Science* (IF: 31.2), *PNAS* (IF: 9.68), and *Nature* (IF: 36.28) have the most retracted articles, while 16 journals with IFs less than 3 (as is the *American Journal of Neuroradiology* [*AJNR*]) had none (in my time as Editor-in-Chief, only 1 *AJNR*-related article had to be retracted, and this was actually done by another journal because the original appearance of the duplicated article was published by us). Because retractions are generally initiated by journal editors and some may not wish to accept the mistake of publishing a fraudulent article, many articles that should be retracted are not and remain viable and gain citations. Thus, the current number of retractions is probably underestimated.

Stephen Breuning was the assistant director of the largest institution for the mentally impaired in Pennsylvania. In 1983, it was discovered that he falsified data presented in a symposium abstract, which led the National Institute of Mental Health to review his publications, reaching a conclusion that 24 of 25 were fraudulent.[2] Surprisingly, only 3 were retracted at that time, and 24 years later, a study showed that they continue to be cited even by prestigious journals such as the *British Journal of Psychiatry* (IF: 6.61).[3] In another study, 235 retracted articles accumulated 2034 citations, and depending on how the data were analyzed, the retractions were acknowledged in only 6.4%–

7.7% of the journals.[4] Also, "infamous" articles may be quoted more often than "famous" ones. On the Scholarly Kitchen Web site* (http://scholarlykitchen.sspnet.org), Kent Anderson said about retractions: "In high impact journals, there is no reason to believe that these citations don't contribute their fair share to the impact factor. After all, an infamous paper may be more readily cited because it's top of mind for a busy author."[5] (This is a common joke among journal editors: if you want your IF to go up, publish a fraudulent article!)

If we have the IF, the *h*-index, and other metrics, why not have a retraction factor? Drs. Ferric Fang and Arturo Casadevall, editors of *Infection and Immunology* (IF: 4.16) and *mBio* (IF: 5.3), respectively, set out to do this. Fang and Casadevall[6] simply took the number of retractions per journal from 2001 to 2010 and divided it by the total number of articles appearing on PubMed during the same period. Because the number of retractions tends to be small, they multiplied their results by 1000 to obtain whole numbers.

This recent article pointed out again that retractions occur more often in higher IF journals. In a different article, the policies on retractions found in major biomedical journals were studied.[7] For this investigation, the author selected the 122 journals with the highest IFs and found that 62% did not have a formal policy regarding retractions (*AJNR* does and it can be found at http:// www.ajnr.org/site/misc/ifora.xhtml#dupl). In August 2012, *The Scientist* published an opinion piece calling for a "transparency index" similar in spirit to the IF.[8] The authors suggested that this index should include the following: the article review protocol of the journal (*AJNR* has one), whether underlying data are made available (*AJNR* does not unless something is called into question), whether the journal uses plagiarism-detection software (yes, *AJNR* does this), whether a mechanism for dealing with fraud issues exists (see the Web address above for the policy of *AJNR*), and whether corrections and retractions are as clear as possible (I believe ours are).

Older research told us that retractions took some time to take place; an observation that no longer holds true as seen in a recent investigation.[9] In that study, the entire universe of biomedical literature between 1972 and 2006 was examined. While other investigations have used loose controls, in this one, the authors chose as controls only articles published immediately before and after a retraction and in the journal where the fraud had occurred.

Let me spend a few lines here because their results were very interesting.

They found that most fraudulent articles were authored by top researchers at US universities and that retracted articles were likely to be highly cited in their first year. However, they also found some good and honest things that happen after retractions:

The system is fast with nearly 50% of retractions occurring 2 years post publication (and it seems that this delay is getting shorter with time), retractions are unbiased, and the effects of retractions are severe and long-lived (citations for retractions were down 72% by 10 years). Most retractions are American articles, and the number of retractions reported by journals published outside the United States is small. Does this imply that research done elsewhere is more honest? I believe that this is not the case and that foreign journals perhaps have less well-established policies and procedures on retractions and/or pay less attention to this problem.

If you want to be entertained (not to say amazed or even disgusted) by the retraction epidemic, I suggest visiting the Retraction Watch Web site (http://retractionwatch.wordpress.com).

This is a moderated blog that reports instances of fraud and allows visitors to comment. Recently posted, there is a new twist: retraction of an article "in press," meaning that its final version was not yet available and that it had not been assigned space (issue, pages) in the journal.[10] The implication is that fraud is occurring and being detected even at the preliminary submission stage. As in many other cases, the reason for this retraction was opaque and listed as "article withdrawn at the request of the authors and editor."

However, all of this should not come as a surprise. An article by John Ioannidis, a meta-researcher who specializes in this sort of thing, states that 80% of nonrandomized studies (the most common type published) are eventually proved wrong as are 25% of randomized trials and 10% of large multi-institutional randomized ones.[11] He has been able to identify the characteristics of articles that make them more likely to contain false information: small populations, small overall effect on science, financial interests, and a "hotter" field, among others. All of these features contribute to eventual retractions. Dr Ioannidis said, "At every step in the process, there is room to distort results, a way to make a stronger claim or to select what is going to be concluded."[12]

As shown in 2 cases just last year, the retraction (and fraud) epidemic continues and is growing. The first case involved the dean of the School of Social and Behavioral Sciences at Tilburg University in the Netherlands.[13] His studies involved the effects of trash-ridden environments and eating meat on behavior, and some were published in *Science*. Dr Stapel's deception was driven (according to him) by his "quest for beauty—instead of truth."

People like him, obsessed by order and symmetry, have difficulty accepting the often messy results of research. A university committee concluded that 55 of his articles were fraudulent, and he is now being investigated for misuse of public funds given to him in the form of grants.

In the second case, Dr Yoshitaka Fujii, a researcher and anesthesiologist from the University of Tsukuba in Japan went on for years publishing fraudulent articles.[14] The first allegations of fraud came in 2000, and by 2012, a panel of investigators had concluded that he had been publishing falsified data since 1993. In April 2012, twenty-three journals publicly requested that the Japanese Society of Anesthesiology investigate Dr Fujii. By June, a commission had found that 172 of his articles contained some fabricated data, and of these, 126 were "totally fabricated."

This last example is the largest case of scientific fraud to date, and I am sure that, unfortunately, it will not be the last. In 1 survey, 2% of academics admitted to falsifying or fabricating data and 28% claimed to know colleagues who had done it.[15]

* The Scholarly Kitchen is a moderated blog established by the Society for Scholarly Publishing to "advance communication through education and networking." It is a must for anyone interested in scientific publication.

REFERENCES

1. Fang FC, Steen RG, Casadevall A. **Misconduct accounts for the majority of retracted scientific publications.** *Proc Natl Acad Sci U S A* 2012;109:17028–33
2. Boffey PM. **US study finds fraud in top researcher's work on mentally retarded.** *The New York Times.* May 24, 1987. http:// www.nytimes.com/1987/05/24/us/us-study-finds-fraud-in-topresearcher-s-work-onmentally-retarded.html?src_pm. Accessed November 5, 2013
3. Korpela KM. **How long does it take for the scientific literature to purge itself of fraudulent material?: The Breuning case revisited.** *Curr Med Res Opin* 2010;26:843–47
4. Budd JM, Sievert ME, Schultz TR, et al. **Effects of article retraction on citation and practice in medicine.** *Bull Med Libr Assoc* 1999; 87:437–43
5. Anderson K. Mountains out of molehills and the search for a retraction index. September 2011. http://scholarlykitchen.sspnet.org/2011/09/01/mountains-out-of-molehills-and-the-search-for-a-retractionindex. Accessed November 5, 2013

6. Fang FC, Casadevall A. **Retracted science and the retraction index.** *Infect Immun* 2011;79:3855–59
7. Atlas MC. **Retraction policies of high-impact biomedical journals.** *J Med Libr Assoc* 2004;92:242–50
8. Marcus A, Oransky I. **Bring on the transparency index.** *The Scientist.* August 1, 2012. http://www.the-scientist.com/?articles.view/ articleNo/ 32427/title/Bring-On-the-Transparency-Index. Accessed November 5, 2013
9. Furman JL, Jensen K, Murray F. **Governing knowledge in the scientific community: exploring the role of retractions in biomedicine.** *Research Policy* 2012;41:276–90
10. Is an "article in press" "published?" A word about Elsevier's withdrawal policy. Retraction Watch. http://retractionwatch.com/2013/02/25/is-an-article-in-press-published-a-word-about-elseviers-withdrawal policy.
Accessed November 5, 2013
11. Ioannidis JPA. **Why most published research findings are false.** *PLoS Med* 2005;2:e124
12. Freedman DH. **Lies, damned lies, and medical science.** *The Atlantic.* November 2010. http://www.theatlantic.com/magazine/archive/2010/11/lies-damned-lies-and-medical-science/308269. Accessed November 5, 2013
13. Bhattachariee Y. **The mind of a con man.** *The New York Times.* April 26, 2013. http://www.nytimes.com/2013/04/28/magazine/diederikstapels- audacious-academic-fraud.html?pagewanted_9&_r_1&src_dayp. Accessed November 5, 2013
14. Yoshitaka Fujii. http://en.wikipedia.org/wiki/Yoshitaka_Fujii. Accessed November 5, 2013
15. Fanelli D.**How many scientists fabricate and falsify research? A systematic review.** PlosOne 2009; 29: e5738

PEER REVIEW: PAST, PRESENT, AND FUTURE

There are basically 3 types of peer review in the field of medicine:

1. Medical peer review, by which organizations are appraised.
2. Clinical peer review, by which skills of physicians are evaluated.
3. Scientific peer review, by which articles submitted to journals are reviewed and their quality assessed.[1]

Here, I will concentrate on the last type of peer review.

Scientific peer review dates back to 1752, when the Royal Society of London established a committee to assess publications submitted to their journal, *Philosophical Transactions* (a practice that may actually have originated in Edinburgh before this event).[2] The journal editor considered this as an "internal peer review," meaning that no individuals working outside of the journal looked at the articles. The purpose of the reviewers was to help the editor choose articles considered as appropriate for the general theme of the journal and not to strictly address their quality. Because the journal had considerable page space, rejections were rare and the goal was to fill it. In the early 1900s, scientific peer review, as we now know it, began in earnest in the United States. The first journals to use it were *Science*, *JAMA*, and *American Practitioner*. Peer review was not practical until 1959, when the photocopier was invented. At that time, multiple copies of submitted articles could be mailed to external reviewers without fear of losing them. During the next decade, the overall activities related to the science of medicine and its research increased dramatically, and the previous excess page space in most journals disappeared.

Thus, editors saw the need to be more discriminating, and peer review began assessing the quality of submissions and only publishing those that passed a rigorous review. This is the type of peer review most current journals use and, in a poll, over 96% of scientists supported it.[3]

Two types of peer review dominate scientific journals:

Closed.

1. Unblinded: the names of authors and reviewers are known to each other; less commonly used than other systems because of the fear of introducing bias in the process.
2. Blinded:
   - Single-blinded: authors know the names of reviewers and may even suggest them (called "author-guided review") or reviewers know the name of the authors (more common than the former and somewhat more controversial).
   - Double-blinded: neither authors nor reviewers know each others' names (used by most journals, including *AJNR*; number of reviewers varies per article between 2 and 3).

Open. Materials are generally posted on the World Wide Web (WWW) and are open to review by all users (even the general public). This may occur before or after publication (vide infra).

The effects of blinding the authors to the reviewers, and vice versa, have been studied. In one such study, keeping the names of the reviewers secret led to 8% more rejections than when they were known to the authors, implying that reviewers are more willing to reject a submission if they know that their identities will not be made known.[4] Conversely, publicizing the reviewers' names led to a 5% increase in positive recommendations.

[5] In an assessment of which type of peer review is preferred, a survey of 838 individuals showed that 68% of reviewers favored not knowing the authors' names and that 72% of authors chose not to know the reviewers' names, findings that support the use of double-blinded peer review.[6]

Dr. Richard Smith, a former editor of the *British Medical Journal*, called peer review "expensive, slow, prone to bias, open to abuse, anti-innovatory, and unable to detect fraud."[7]

In addition, the following problems have been noted with scientific peer review[8]:

1. Blinding reviewers to the authors' identities does not improve the quality of evaluations.
2. Passing reviewers' comments to their co-reviewers has no effect on improving the quality of future reviews.
3. Spending more than 3 hours doing a review does not increase its quality.
4. Measuring the quality of peer review is challenging.

Who is a good reviewer? According to one study, the best manuscript reviewers are individuals younger than 40 years of age (I would have thought that the more experienced, and thus older reviewers, were better, but that is not the case), from top academic institutions (makes sense to me), personally known to the editor (I call this the "shame factor"), and those who are blinded to the identity of authors.[9] If an individual has all 4 characteristics, 87% of his or her reviews will be judged as being excellent. If, however, the reviewer possesses only 1 characteristic, only 7% of his or her reviews will be excellent.

In a different study, aimed at evaluating how carefully the reviewers analyzed submissions, the editors introduced "8 areas of weakness" (read: errors) into 1 article and sent it out to 420 reviewers.[10] Of them, 53% completed the review, but only an average of 2 errors was detected by all. The investigators concluded that "neither blinding reviewers to authors and origin of the paper nor requiring them to sign their reports had any effect on the rate of detection of errors. Such measures are unlikely to improve the quality of peer review reports." Another article's goal was to investigate which errors were detected by reviewers and if prior training in reviewing improved their ability to spot these.[11] Nine errors were inserted into 3 articles and given to trained and untrained reviewers.

Overall, only 2–3 errors were detected, with biased randomization being the most frequently recognized. Training the reviewers did not significantly improve their reviews.

As alluded to before, misconduct also occurs in peer review in the form of personal vendettas, abuse of anonymity, false praise of submissions due to fear of vindictive authors, and, more importantly, plagiarism of unpublished data for personal benefit.[12] Dr. J. Eisen, editor of *PlosONEBiology*, had the following to say about peer review: "If you asked someone today to design from scratch a peer review system, they would not design it the way it is," and "Having 2 or 3 reviewers and one editor as gatekeepers of scientific knowledge is a mistake.

It has too much potential for limiting the spread of scientific knowledge." In 2008, the Cochrane report stated the following[13]:

1. There is no clear effect of author or reviewer blinding.
2. There is no evidence that reviewer training improves the quality of the process.

3. Different ways of communicating with reviewers have no effect on the quality of reviews.
4. There are little data to support that peer review improves the quality of published articles.

Thus, the next question that comes to mind: Is there a better process? The quest continues and today "open" peer review is receiving a lot of attention. As peer review was only possible after photocopies were widely available, open review was only possible after the WWW matured and morphed into Web 2.0. Benefits of open review are increased number of reviewers, increased transparency, more constructive criticisms (the last 2 observations are based on the fact that the names of reviewers are generally known and that all discussions remain archived on the Web), and higher quality submissions. [14] Drawbacks include that sometimes it is not possible to get enough reviewers (because it is a completely voluntary and unsolicited activity), delayed publication, and, overall, a system that is more complex than double-blinded peer review. There are 2 types of open reviews: pre- and post publication.

A detailed explanation as to how each type works would be too long for this *Perspectives*, and I have included these in pictorial form (Fig 1). While some journals have successfully implemented open review, others have failed. In 2006, *Nature* allowed authors to choose between their traditional peer-review system and open review, and only 71 of 3000 opted for the latter.[15] At the end of the trial, the editor commented as follows: "From informal feedback, it was clear that the trial generated a lot of casual interest, but no hostility or enthusiastic endorsements in any quantity. Unsolicited comments posted on the Web were less useful than those from designated referees but in principle could draw attention to something not spotted by the referees."

How about a hybrid peer-review system? This implies open review for only selected articles. The *Proceedings of the National Academy of Sciences* uses such an approach and open reviews occur after acceptance and posting of articles. Open reviewers like it because they are not responsible for acceptance/ rejection, and the assigned editor and original referees remain anonymous. This is true of post publication peer review, and even journals outside of the sciences, such as *Shakespeare Quarterly*, have successfully implemented it. An easy way to go about post publication open review is to use blogs and social media. Benefits include openness, keeping articles fresh by continuous evaluations and changes, and publication schedules that are not significantly affected. Drawbacks are amateurish evaluations and the fact that people are reluctant to blog (similar to our experience with ajnrblog.org).[16] *Plos ONE* has used post publication open peer review successfully.[17]

Mr. Viter Tracz (responsible for Gower Medical Publishers, the *Current Opinions Journals*, and BioMed Central) considers himself the father of open access and said the following of peer review: "Except for a tiny little part at the top where it is done seriously, peer review has become a joke. It is not done properly, it delays publication unnecessarily, it is open to abuse, and is being abused. It is seriously sick, and has been for a while." He believes that a large group of volunteer scientists should do open review and thus created the "Faculty of 1000."[18] This Web site specializes in post publication rating of articles and contains over 112,000 as of this writing. The neuroimaging section of F1000 is mostly composed of basic science, but the articles rated include clinical ones, too. Articles are rated on a numeric scale and given comments such as "Must Read," "Exceptional," and so on. They also rank journals (*AJNR* is ranked 813 of 1129) based on the number of articles F1000 reviews, article grades, and total yearly publications of that journal. So, they ranked *AJNR* based on only 2 articles! Other similar Websites, such as Evidence-Based Medicine, use a star rating system a la Amazon (this type of rating became available in *AJNR* in March 2012). On both sites, the articles tend to be older, as they only rate those that are open access (1 year for *AJNR* but 2 years for most other journals).

The *Radiology Best Evidence Newsletter* from Medscape is more contemporary, but because most articles rated are not open access, only their abstracts are found there.[19] Facebook may also be used as a means of post publication open peer review, and the *New England Journal of Medicine* has been successful with this method. Conversely, *AJNR* has not, but maybe as neuroradiologists become younger, it will.

Whatever peer review system we continue to use, we need to be careful, as governments are starting to look into this issue. In England, the House of Commons Science and Technology Committee issued a report on "Peer Review in Scientific Publications."[20] They concluded that there are many ways of doing peer review, that publishers should offer a variety to suit the needs of different publications, and that the importance of prepublication assessment is crucial and this always requires subjective judgments that may result in errors. They encouraged different research groups to optimize review systems and foster innovations, and stated that openness and transparency are attractive, and, at the end, they congratulated *Plos ONE* on the quality of their on-line programs. More importantly (although not directly related to the topic of this *Perspectives* but one that is the result of peer review), they stated that the use of the Impact Factor to measure the quality of articles is questionable (I imagine this is where the importance previously placed on post publication review comes in) and that the Impact Factor should not be used when assessing individuals for career progression.

To end on positive note, a former editor of *JAMA* said, "Peer review represents a crucial democratization of the editorial process; incorporating and educating large numbers of the scientific community, and lessening the impression that editorial decisions are arbitrary."[21]

AJNR uses the time-honored double-blinded peer-review system, but our readers and the general public are welcome and encouraged to use our blogsite and Facebook page as a means of post publication open review.

References

1. http://teachingcommons.cdl.edu/cdip/facultyresearch/Historyofpeer review. html. Accessed on December 19, 2011
2. Spier R. **The history of the peer-review process.** *Trends in Biotech* 2002;20:357–78
3. http://www.highbeam.com/doc/1G1–139473352.html. Accessed on December 19, 2011
4. Schroter S, Tite L, Hutchings A, et al. **Difference in review quality and recommendations for publication between peer reviewers suggested by authors and editors.** *JAMA* 2006;295:314–17
5. van Rooyen S, Godlee F, Smith R, et al. **Effect of open peer review on quality of reviews and on reviewer's recommendations: a randomized trial.** *BMJ* 1999;318:23–27
6. Regehr G, Bordage G. **To blind or not to blind? What authors and reviewers prefer.** *Med Educ* 2006;40:832–39
7. Smith R. **Opening up BMJ peer review. A beginning that should lead to complete transparency.** *BMJ* 1999;318:4

8. Goldbeck-Wood S. **Evidence on peer review—scientific quality control or smokescreen?** *BMJ* 1999;318:44–45
9. Estrada C, Kalet A, Smith K, et al. **How to be an outstanding reviewer for the Journal of General Internal Medicine...and other journals.** *J Gen Intern Med* 2006;1:281–84
10. Godlee F, Gale CR, Martyn CN. **Effect on the quality of peer review of blinding reviewers and asking them to sign their reports. A randomized controlled trial.** *JAMA* 1998;280:237–40
11. Schroter S, Black N, Evans S, et al. **What errors do peer reviewers detect, and does training improve their ability to detect them.** *JRSM* 2008;101:507–14
12. http://www.genomeweb.com/peer-review-broken. Accessed on December 19, 2011
13. http://srdta.cochrane.org/sites/srdta.cochrane.org/files/uploads/Editorial_Process_for_DTARS_web_document.doc/5B1/5D.pdf. Accessed on December 19, 2011
14. http://www.atmospheric-chemistry-and-physics.net/pr_acp_peer_review_ steps_out_of_the_shadows.pdf. Accessed on December 19, 2011
15. http://www.nature.com/nature/debates/e-access/index.html. Accessed on December 19, 2011
16. http://www.progressivehistorians.com/2008/01/blogging-and-peer-review.html. Accessed on December 19, 2011
17. http://www.plosone.org/static/guidelines.action#production. Accessed on December 19, 2011
18. http://f1000.com. Accessed on December 19, 2011
19. http://www.medscape.com. Accessed on December 19, 2011
20. http://www.publications.parliament.uk/pa/cm201012/cmselect/cmsctech/856/ 856.pdf. Accessed on December 19, 2011
21. Rennie D. **Fourth International Congress on Peer Review in Biomedical Publication**. *JAMA* 2002;287:2759–60

Fig 1. Diagram comparing 3 different types of scientific journal peer review. On top (*white boxes*) is our traditional double-blinded peer review. In the center (*pink-shaded boxes*) is a hybrid system that incorporates external open review into the double-blinded one. This results in a time penalty and increased expenses and thus is not practical. On the bottom (*red boxes*) is an open-review-only system that may save time and expenses in publication but remains controversial.

GLOBISH AND THE EMPIRE

When an empire collapses, one of the most important legacies it can leave behind is its language. Spanish is the fourth most commonly spoken language in the world. Nearly 400 million people do so, most in Latin America. The Spanish Empire started to gain force in the early 1500s and attained its peak with the colonization of the Caribbean and the Americas.

[1] By the time the empire had nearly disappeared in the mid-1800s, Spanish was the official language in more than 20 countries. To be sure, the Spanish spoken outside of Spain has mutated. My

wife often reminds me that I do not speak Spanish, but "Caribbean." Perhaps, only in distant Argentina and Chile does Spanish remain closest to its roots. Curiously, where Spanish has mutated, many of its original expressions remain intact. My wife also points out that at times I tend to speak like her grandmother—that is, many of the idiomatic phrases of long ago remain frozen outside of Spain. Of course, Spanish, like other languages, has become "tainted" by English, something that the Real Academia de la Lengua Española battles with constantly.

The French Empire was different from the Spanish one in that its expansion occurred not only far away but also within Continental Europe.[2] There are currently 21 Francophone countries outside of Europe and 4 in it. To the *chagrin* of the French, these countries are small except for Canada, the Republic of Congo, and the Democratic Republic of Congo, and French ranks eleventh in a list of the most commonly spoken languages (about 129 million individuals).[3] The diffusion of the French language is interesting because in the 19th and early 20th centuries, it was the language of culture, science, politics, and diplomacy and was studied and spoken by all learned individuals.

As the French will admit—and not so happily— their language has also become "tainted" by *anglicismes*; witness:"le weekend," "le shopping," "le parking," etc.

> Despite efforts of the Academie Francaise, synonyms for many English words do not exist in French. In the largest Francophone cities—Kinshasa, Brazzaville, and Montreal—French has also strayed from its roots.

After English, perhaps the most successfully exported language is Portuguese. Outside of Portugal, there are 7 countries where

Portuguese is the official language. It ranks seventh in the world's most commonly spoken languages and is the second most common in South America. About 250 million individuals speak it worldwide. The Portuguese Empire started about the same time as the Spanish one but ended earlier, in the early 1700s, when many of its territories were taken over by the Dutch and its trading along the Indian Ocean came to an end.[4]

Languages may also spread not only on the basis of actual physical colonization but by economic factors. By virtue of its large size and growing economy, China's languages (especially Mandarin) are now the second most commonly learned after English.[5] China is clearly becoming the great economic empire of the 21st century, and Mandarin is spoken by more than 850 million people, followed distantly by English (510 million) and Hindustani (490 million). Despite this, Spanish, Portuguese, and French are still highly ranked as popular languages to learn.

One of the fiercest colonizers were the British. Their empire started late by European standards—16th and 17th centuries—but by the early 1900s, they dominated more than one-fourth of the world. Their most important outpost was the original 13

American colonies, from where English would propagate to the rest of the world. The end of World War II brought to a close the last vestiges of that great empire but led to a tremendous expansion in the use and learning of its language.

There is no doubt that the most successful export of the United States has been its culture and language. Although we do not speak about an "American Empire," we certainly speak of "American Imperialism." This last term reflects the military, economic, political, and cultural influence of our society on the rest of the world.[6] American imperialism began in the late 1800s after victory in the Spanish-American War and still goes on today. Even though our economic status is declining, globalization has assured that our culture and language continue to be omnipresent. Everywhere you go, you find American fast food, for which we do not want to be known. However, all places you go you can speak English, and our scientific status continues to be admired.

Even if the rest of the world seems intent on learning English, we Americans speak few languages. The US State Department has divided languages according to their ease of learning.

One can achieve minimal proficiency with 600 hours of learning in the Latin and Germanic languages (so-called category 1 languages). To be minimally proficient in a category 2 language, you will need about 1100 hours of class work (these include Slavic, Turk, other Indo-European languages such as Persian and Hindi, and some non-Indo-European such as Georgian, Hebrew, and many African languages; Swahili is ranked easier than the rest, at 900 hours). The hardest (category 3), requiring more than 2200 hours of learning, are Arabic, Japanese, Korean, and the Chinese languages. In the United States, less than 9% of the population is bilingual; multilinguality is even less common. We are not alone. Although Anglophones are generally monoglots, Hispanophones and Francophones are also guilty of uniligualism.

Whereas monolinguality is limiting, worse is semilinguality.

Children of immigrants face this significant problem— that is, before correctly learning their native language, a second is forced upon them, resulting in suboptimal command of both. If you happen to learn a second language before puberty, you will probably be a "compound bilingual." In this situation, words in different languages are anchored by similar abstract concepts. This means your communication competency will be high in both languages. When it comes to English and Spanish, I consider myself a compound linguist. It really does not matter if my PowerPoint presentations are written in either language or in which language I must give them. In this scenario, languages have truly become communication tools and may be easily interchanged. As one starts to learn another language, the new one is said to be *subordinated* to the first one.

Practice will eventually result in the new language becoming *coordinated* with the maternal one. However, even coordinated speakers will use their original language to think through the newly acquired one. This is the reason why subordinated or coordinated speakers have lower communication competencies in their new language. Languages are currencies and make us richer and more prestigious. Polylinguality is not democratic; one language is generally more prestigious than others. It is not uncommon for the heritage (maternal) language to be replaced, generally in the second generation, by a more prestigious learned one (a common situation in the United States). Today, there is no more prestigious language than English. Not only is it the universal language of science, but it is the true lingua franca.

During the first few years of the second millennium, it seemed the fate of English was in doubt. The unpopularity of American foreign policy seemed to threaten a widespread acceptance of English. Then, suddenly, Barack Obama was elected, and American English recovered its prestige as the language of democracy and diplomacy. Curiously, by the time this happened, the English spoken throughout the world had changed and was on its way to become what is now known as Globish. The term "Globish" was coined by Jean-Paul Nerriere in 1995.[7] Globish initially comprised about 1500 words that were mainly intended for business transactions, but as Robert McCrum tells us in his fascinating book, *Globish: How the English Language Became the World's*

Language,[8] Globish is like a river accepting many tributaries and has become enormous and laden with many new words. One can think of Pidgin English as an ancestor of Globish. This simplified version of English served a purpose when different groups of people needed a common ground. African slaves forced into the Americas would not have survived in its absence. Historically, "black" English has been extremely mutable due to its predominant oral transmission (as is Globish). The most recent iteration of this is African American vernacular English (Ebonics), whereas the most recent iteration of English is Globish.

Words like "hip," "cool," "cat," and "chick" are now parts of everyday English (and of Globish). Globish speaks to all of us, and Mr. Obama is a true master of a universal English ("I learned to slip back and forth between my black and white words."—*Dreams of My Father*[9]). Obama's English (and Globish) is perfect (and elegant) for our globalized world. Everything about a globalized world—literature, science, Internet, economy, mass tourism—points toward the need for a common language. Globish will certainly rid us of the anxiety of miscommunication so pointedly addressed by Sofia Coppola in the film *Lost in Translation*.

Although speaking with the correct accent is desirable and graceful, Globish pays little attention to this. As English continues to be universally adopted, accents become less critical and acquire a certain importance because they inform us of the speaker's background and nationality. In America, we continue to be somewhat intolerant of accents. Proof of this is the countless complaints at scientific meetings when non-native speakers have the courage to use English. How many of us would be brave enough to present in Mandarin in front of a Chinese audience? We should remember that English has become stronger thanks to China. The Chinese movie *Crazy English* follows Li Yang throughout his English teachings. His motto, "conquer English to make China stronger," leads me to believe that once the Chinese indeed conquer it, English will truly be the real lingua franca.[10] The second engine in reinforcing the use of English as the lingua franca is India, for its high-tech centers, such as Bangalore and its impressive entertainment industry (Bollywood), rely on it.

About 5000 languages are now spoken in the world. Within the next century, more than 90% of these will disappear.[11] The Bible says: "If as one people speaking the same language they have begun to do this, then nothing they plan to do will be impossible for them" (Genesis 11:6). Unfortunately, this follows:

"Come, let us go down and confuse their language so they will not understand each other." Attempts to establish universal languages have failed before (who of us can speak Esperanto?). Latin, once the lingua franca, split into many languages that today share little with each other. Let's try to use English to improve ourselves and our sciences, for science may be a new empire and only a unified one can be successful and lasting. At the *American Journal of Neuroradiology*, we take the usage of English seriously, but we also recognize that our English- as-a-second-language authors comprise most of our contributors.

My advice for them is straightforward: There is no way to avoid Globish, so keep it simple and do your best not to confuse your colleagues.

References

1. History World. http://www.historyworld.net. Accessed August 3, 2010
2. Wikipedia. First French empire. http://en.wikipedia.org/wiki/First_French_Empire. Accessed August 3, 2010

3. Listverse. Top 10 most spoken languages in the world. http://listverse.com/2008/ 06/26/top-10-most-spoken-languages-in-the-world. Accessed August 3, 2010
4. Wikipedia. Portuguese empire. http://en.wikipedia.org/wiki/Portuguese_Empire. Accessed August 3, 2010
5. Ezine @rticles. Learn Chinese language: the worlds most frequently learned language after English. http://ezinearticles.com/?Learn-Chinese-Language—The-Worlds-Most-Frequently-Learned-Language-After-English&id_591081. Accessed August 3, 2010
6. Wikipedia. American imperialism. http://en.wikipedia.org/wiki/American_ imperialism. Accessed August 3, 2010
7. Don't speak English. Parlez globish. http://www.jpn-globish.com. Accessed August 3, 2010
8. McCrum R. *Globish: How the English Language Became the World's Language*. London, UK: Norton; 2010
9. Obama B. *Dreams of My Father.* New York: Crown; 2007
10. Osnos E. The national scramble to learn a new language before the Olympics. Letter from China. Crazy English. *The New Yorker.* http://www.newyorker.com/ reporting/2008/04/28/080428fa_fact_osnos. Accessed August 3, 2010
11. Nettle D, Romaine S. *Vanishing Voices: The Extinction of the World's Languages*. Oxford University Press, 2002

PRESERVATION OF KNOWLEDGE, PART 1: PAPER AND MICROFILM

In his wonderful, award-winning book *Double Fold*, Nicholson Baker makes a plea for preservation of printed materials, particularly newspapers.[1] He reports on the constant destruction of printed materials that have been preserved on microfilm, the latter presumably being more resilient to the passage of time than paper itself. Historically, paper was made from plants with long fibers (such as linen, cotton, and esparto), yielding a high-quality and long-lasting product. Some of the best linen paper, made in Japan, can survive for centuries. In a similar fashion, cotton-based paper is also of very high quality and long lasting. With the ever-increasing popularity of the printed word, new and more efficient ways of manufacturing paper were sought, and at approximately the mid-1800s it was discovered that wood pulp could be used for this purpose.[2]

The Swedish were the first to use wood pulp for paper on an industrial basis.

Plants and trees, the main source of pulp for papermaking, also contain important polymers called lignins. These serve a function in the transport of water and in giving plants their structural strength. Here is an analogy: lignins are to plants as concrete is to a brick wall. If we grind a tree, the resulting pulp will contain lignin and fibers. To make stiff and strong paper (such as that found in supermarket bags), one needs a lot of lignins. The problem is, with time, lignins deteriorate and give off acids (particularly carboxylic ones). These acids cause paper to turn yellow (this is no problem with brown paper bags and other dark products with a short life). Exposure to light hastens this process, as does exposure to ambient air (comic book collectors tend to keep them in plastic bags). The byproducts of lignin deterioration damage cellulose, and the paper becomes brittle. That is why old newspapers are yellow-tobrown and tend to break off at creases.

One way to remove lignins is bleaching. Once the pulp has been extracted, it can be treated with a mild base (usually calcium or magnesium bicarbonate) to neutralize the naturally occurring acids. Additional base may be added to prevent acidity brought on by the process of "sizing." Sizing generally involves the application of an acidic polymer to the surface of the paper that will basically prevent ink from spreading too much in the paper fibers (paper for color laser printing is heavily sized, making it is difficult to write on with a fountain pen, as the ink will not be absorbed and takes a long time to dry). Extraction and neutralizing of lignins make paper white (brown paper bags may contain up to 55% of unbleached pulp).

The process of bleaching builds into paper the so-called alkaline reserve. For paper to last at least 100 years, its alkaline reserve needs to be approximately 2%. Alkaline paper (called acid-free) can survive anywhere from 500 to 1000 years depending on its quality. Alkaline paper has innumerable advantages vs traditional acid paper including less wear on the machinery that produces it, papermaking by-products that are recyclable, less energy needed to dry it, and the paper itself being easier to recycle. In the paradoxic sense, as we add more recycled paper into new, we are increasing the content of postconsumer fibers and, again, introducing lignins and acidity.

For completeness' sake, I would like to note that the absolute whiteness of the paper we commonly use cannot be achieved solely by bleaching. Initially, the pulp is thermo-treated (basically cooked), and whatever lignin is left behind from this process is then removed by bleaching. Bleaching initially involved the use of chlorine, but because of environmental concerns, other chemicals are now used. Thus, the first step in preservation of printed materials is improvement of the quality of paper.

As early as 1930, librarian William J. Barrow noted the deterioration of paper publications. He wrote several seminal articles on this topic from the 1940s to the 1960s and headed an important laboratory concerned with investigations regarding the quality of paper. Barrow's initial observations went basically unheard until the Council on Library Resources and the American Library

Association granted him monies to pursue his investigations. In 1988, the National Endowment for the Humanities began the Brittle Books Program, which involved the microfilming of 3 million decaying books. The program also includes the deacidification of books that are still in good condition (because this is an aggressive process, it will actually damage those books that are decaying).

Microfilming, as evidenced by Mr. Baker's book, is highly controversial. Microfilm can last for approximately 500 years (less than high-quality paper), but it needs to be stored in proper conditions and viewed with special machines. Unfortunately, the quality of the material archived on microfilm is highly variable and many times is unreadable because sufficient care was not taken to optimally photograph the original publications. In 1984, the National Information Standards Organization (www.niso.org) proposed voluntary standards for paper manufacturing that included pH value (the pH of alkaline paper is 6.0 –7.0), tear resistance, alkaline reserve, and lignin concentration. If followed, paper complying with these suggestions may last thousands of years. In 1994, the International Standardization Organization (www.iso.org) proposed international equivalent standards. Furthermore, both organizations suggest that publications of special significance be printed on archival-grade paper. The highest quality of this type of paper is called museum grade and is cotton based.

In 1987, the National Library of Medicine (NLM) created the Permanent Paper Task Force.[3] This institution encourages the use of permanent paper that meets the criteria outlined in the Permanence of Paper for Publications and Documents in Libraries and Archives Act; all journals must be printed on acid-free paper and are identified as such in PubMed. Our printer, Cadmus, uses acid-free paper for the cover and contents of the *American*

Journal of Neuroradiology (*AJNR*) and meets the requirements of the American National Standards Institute (www.ansi.org). The mill that produces our paper uses environmentally friendly chlorine-free bleaching.

So, although the issue of paper preservation has been at least partially solved, the problem regarding the amount of space required to store scientific journals has not. Of course, archiving on microfilm and microfiche has not proved to be as easy or reliable as initially thought. The NLM has a section dedicated to preservation and collection management that deals with these issues. Storage requires controlled temperature, humidity, light, and shelving; transportation is delicate and tricky, and viewing exposes film to significant physical stress that contributes to its deterioration. Microfilm is a cellulose acetate-based product that is very sensitive to physical trauma. Most university libraries own equipment that will allow one to print, e-mail, save images to USB devices, or burn them on CD or DVD. Loading the film tapes into these machines can be risky, and specific instructions need to be followed.

Digitization of microfilm is being done but requires scanners capable of resolutions close to 10,000 dots per inch.

The one thing microfilm has clearly achieved is space savings.

Storage requirements are reduced by nearly 95%. It also prevents further deterioration of original manuscripts by avoiding repeated handling. Color microfilm is extremely expensive; thus, most archives are only in black and white. Why the NLM chose microfilm to archive its material has been addressed in many articles and books. The development of microfilm during the First World War was related to espionage activities. Before becoming the NLM, this repository was called the Army Medical Library, and it was not until the mid-1950s that the Department of Defense transferred it to the Public Health Service. I doubt many neuroradiologists have consulted the microfiche carriage in our local library lately, as most biomedical data are now stored electronically. Next issue, I will continue this *Perspectives* with a brief account of digital storage activities as they relate to the sciences and to *AJNR*.

References

1. Baker N. *Double Fold: Libraries & the Assault on Paper.* New York: Random House; 2001
2. Lister D. The Lister List. Acid Paper. Available at: http://www.britishorigami. info/academic/lister/acidpaper.php. Accessed March 25, 2009
3. Preservation and Collection Management Section, National Library of Medicine. Acid-Free Paper for Biomedical Literature. Available at: http://www.nlm. nih.gov/pubs/factsheets/acidfree.html. Accessed March 25, 2009

PRESERVATION OF KNOWLEDGE, PART 2: DIGITAL ARCHIVES

In last month's *Perspectives*, I addressed the role of paper and microfilm as media for the preservation of written knowledge, particularly biomedical literature. A significant number of journals are now electronically archived including the *American Journal of Neuroradiology* (*AJNR*). The HighWire Press data base contains 1232 journals and more than 5.6 million articles as of this writing. According to my last count, the Elsevier and Springer journal data bases contain 2320 and 2084 scientific journals, respectively! It is hard to believe all of these journals can be kept in their past and current print forms and archived for posterity. Most current journals are, however, archived in digital form. The main topic of this article is to discuss some of these electronic data bases.

Although one of the major goals of libraries is to avoid duplicated material and save space, one of the primary goals of electronic depositories is to create redundancy and several copies of their contents. The *AJNR* is preserved by a system called LOCKSS (Lots of Copies Keep Stuff Safe).[1] This system is based on technology designed by the Association of Computing Machinery. The LOCKSS system is, like the HighWire Press, based at Stanford University and provides libraries with open-source software to preserve all sorts of materials that have been published on the World Wide Web. Each library owns a LOCKSS "box" that allows perpetual access to their materials. LOCKSS has a self-checking mechanism that continuously audits and repairs the information it houses. This is accomplished by "crawlers" that compare the LOCKSS box contents with an institution's Website and constantly update the box contents.

The *AJNR* allows these crawlers on our Website where information is collected and used to update our LOCKSS files.

However, unlike libraries, we do not administer this mechanism;

HighWire does. Decentralization of the system assures its independence from central failures by creating multiple archives in different locations and constantly comparing these with replicas. Today, approximately 400 publishers and hundreds of libraries worldwide use LOCKSS. The next iteration of this system is CLOCKSS (the C stands for "controlled").

Through it, the main research libraries in the world and major publishers are creating a "dark" archive of all of their contents to further assure the preservation of digital data. LOCKSS can also be used to keep older literature that has been digitized.

Because software and hardware are in a state of constant evolution, data may "rot," and all preservation systems need to be constantly updated.

In June 2004, the Wellcome Trust, Joint Information Systems Committee, and the US National Library of Medicine announced a joint effort to digitize back files of what they considered to be important medical journals.[2] This effort is known as the Medical Journals Backfiles Digitisation Project and is one of a total of 6 such projects in the world. All of the digitized files, some dating back 125 years, became open access.

This data base does not contain any US-based imaging related journals, but a search revealed that the *Korean Journal of Radiology* is included.[3] In addition, some radiation oncology journals are found there.

Digital archiving initiatives are vast. A complete list may be found at the National Digital Information Infrastructure and Preservation Program of the Library of Congress.[4] Other countries have also joined in this effort. The National Library of the Netherlands (KB) has created an "e-depot" system available to all publishers whose main goal is to maintain the integrity of digitally stored objects. This depository is neither "dark" nor "light," but each user has access only as established by a previous individual agreement. All publications contained in KB's e-depot that come from BioMed Central continue to be open access. The storage capacity of this site is expected to reach 1.5 petabytes soon. This endeavor is closely associated with other European ones such as Preservation and Long-Term Access through Networked Services (www. planets-project.eu), the purpose of which is to build practical services and tools to ensure access to digital culture and scientific assets. Another interesting and more far-reaching project is Cultural, Artistic and Scientific Knowledge for Preservation, Access and Retrieval (www.casparpreserves.eu), which preserves all kinds of digital data in a technology-neutral, domain- independent centralized system to assure its longevity.

In the United States, another huge project is Portico (www. portico.org), which is sponsored by the Library of Congress and other nonprofit organizations. The Portico Website lists some interesting facts stressing the importance of digital archiving as follows:

- A total of 13% of articles cited by the *New England Journal of Medicine, Journal of the American Medical Association,* and *Science* are irretrievable from the original hyperlink only 27 months after publication; hence archival of all materials is important as their on-line longevity may be very short.[5]
- In 2002, a total of 70% of faculty in a survey were using electronic journals for research, and 1 year later, nearly 80% considered these as "invaluable research tools."[6]
- Like other digital archiving initiatives, Portico follows the standards set by the National Library of Medicine Journal Archiving and Interchange Suite.[7] These standards set the way documents are saved and transferred. Portico houses information from 487 libraries and 58 publishers (including Elsevier and Springer), including more than 11 million articles. Many other digital-archiving initiatives (JSTOR [Journal STORage at the Mellon Foundation], Ithaka, Aluka, *Journal of the American Medical Association* and *Archives* Journals Backfiles, etc) are available, but I cannot mention all of them in this *Perspectives* because of space limitations.

Digital archiving extends beyond science to all of our daily activities. From the movie industry to our own home videos and photographs, from the nation's digital memory to that of persons, digital archiving is of significant importance. Efuturists assure us that, in the future, all written material will be found in an electronic form and probably will be carried on portable devices. Other imaging-related journals (eg, *Radiol- ogy* and the *American Journal of Roentgenology*) use HighWire and entrust their archives to LOCKSS. Our readers can be assured that the digital contents of *AJNR* are being adequately preserved for future generations.

References

1. **LOCKSS.** Available at: http://www.lockss.org. Accessed March 30, 2009
2. Mehnert R, Cravedi K. **International Agreement to Expand PubMed Central. USLibrary of Medicine, National Institutes of Health.** Available at: http://www. nlm.nih.gov/news/press_releases/intlpubmed04.html. Accessed March 30, 2009
3. **PubMed Central Journals.** Available at: http://www.pubmedcentral.nih.gov/ fprender.fcgi. Accessed March 30, 2009
4. **Digital preservation.** Available at: http://www.digitalpreservation.gov. Accessed on March 30, 2009
5. Dellavalle RP, Hester RJ, Heilig LF, et al. **Information science: going, going, gone: lost internet references.** *Science* 2003;302:787–88
6. Portico. **Why Archive?** Available at: http://www.portico.org/about/why.html. Accessed on April 6, 2009
7. **National Library of Medicine Journal Archiving and Interchange suite.** Available at: http://dtd.nlm.nih.gov. Accessed March 30, 2009

EDITOR'S NITPICKING

For many editors, the "After Deadline" column of the *New York Times* is a must read. In this section, one finds a critique of language, usage, and style. In a similar fashion, I will try to address some language and punctuation mark usage that seems to often confuse our authors. In an era when most of our published articles come from non-native English language speakers, I have found that the items discussed below are often the source of confusion and misuse.

"i.e." versus "e.g."

No, "i.e." does not mean Internet Explorer. In writing, i.e. is meant as an abbreviation for the Latin phrase *id est*. If you are not sure when to use i.e., it is better to say "that is." I.e. also means "in other words" or "it is." I.e. is used to specify or make an expression more clear. I.e. can be used to clarify a preceding statement by restating the idea. Although in the last sentences I have capitalized the expression (because it started the sentences), i.e. (as well as "e.g.") is more commonly found in the middle of sentences and does not need to be capitalized. "I.e." actually looks weird, doesn't it? Similarly, both expressions are so well known that *italicization* is no longer used (it was, long ago). Because both expressions are located midsentence, they tend to be framed by commas, for example, "Eurasia is a large land mass encompassing 2 continents, i.e., Europe and Asia."

"E.g." means *exempli gratis* (a free example of what you are talking about). E.g. is used when you want to clarify a thought by providing the reader with an example. An important thing to remember about e.g. is that you are giving some examples, not all of them! Whatever you list after e.g. is assumed to be only partial, and thus "etc." should not be used after it. Here is a phrase using e.g.: "Europe is a large continent formed by many countries (e.g., France, Spain, Germany, and Italy)." Note that in this sentence, e.g. is found at the end and between parentheses. When located in the middle of a sentence, it may look like this: "Europe is formed by many countries, e.g., France, Spain, Germany, and Italy, which make its population very diverse." Still confused? Then forget the Latin phrases and use only "that is" and "for example"!

"Affect" versus "Effect"

A single letter can make a world of difference. "Affect" (a verb) generally means "to influence." Here is an example: "Current chemotherapy agents affect the way post contrast MR images look." This is the most common scientific usage of the word "affect." With respect to human behavior, it can also be used to mean "acting in a way you do not feel." We have all been guilty of affecting a behavior when we feel threatened, for example, "He affected an air of nonchalance despite knowing that he was wrong."

As used in scientific writing, "effect" (a noun) generally signifies "the result" of something as in "The effect of the drug is to diminish tumor vascularity." If all of this sounds too easy, let's make it a bit more confusing. "Affect" may occasionally become a noun when you are referring to someone's mood, for example, "This drug results in patients displaying a happy affect." "Effect" may become a verb when it is used to mean that something has been accomplished, for example, "The new handwashing regulations should effect a positive change in infection rates."

"Solely," "Only," and "Merely"

"Solely" (an adverb) means "only," that is, completely, entirely, by itself, and without another or others. "Exclusively"(also an adverb) means the same and may be used in the place of "solely" or "only." "Merely" (an adverb) can also mean "only," but it is mostly used to mean "nothing more than," for example, "His posturing is merely attention-seeking." Take the "ly" ending out and you have the adjective "mere" meaning "pure or absolute" or "nothing more than what is being specified" as in "I only got paid a mere $50.00 to read that complicated MR imaging study."

"Neither," "Either," "Or," and "Nor"

"Neither" is a conjunction generally used with "nor" but never with "or," for example, "Neither Paul nor Ringo would have guessed that John really liked them." "Neither" generally is used when referring to 2 things but nowadays is also used when referring to more than 2 items as in "At the end nothing helped, neither courage, strength, morals, nor religion." "Or" is also a conjunction meaning an alternative, "This shirt comes in red or white"; the equivalent of the substitutive character of 2 words; or an approximation, "I will see you again in 3 or 4 days." "Nor" is another conjunction used to negate a clause, a phrase, or words, generally after using "neither." It is a contraction of the Old English word "nother," which when used with "neither" sounds awkward.

"Either" is a positive adjective that is only paired with "or" (never with "nor"). It means being one or the other of 2. It can be used to express a noun or pronoun doing a different thing than the second noun/pronoun, for example, "He wants to buy either a 1.5T or a 3T MR unit."

Remember your verb concordance when using "neither and nor" and "either and or," that is, for singular or plural elements (subjects) in your sentence, the verb must correspondingly be singular or plural.

"Bases" versus "Basis"

"Bases" is the plural of "base" (a noun). It is defined as the foundation, the main ingredient, the starting point, and, in some cases, may be used to imply the groundwork done. In chemistry, it is generally used to denote a liquid solution with a pH greater than 7. In genetics, it signifies any of the 5 purine or pyrimidine bases that form DNA and RNA. It is very commonly used in baseball. When something is "off base," we mean that it is mistaken or wrong. "Basis" means the exact same thing! Why 2 different forms? "Bases" is the *inflected form* of "basis." In the case of nouns, inflected forms are accomplished by adding "s" or "es" to their endings.

Although inflected forms originated with the purpose of giving different grammatic functions to a word or words in a sentence, "bases" and "basis" can generally be interchanged.

"Entirely" versus "Mostly"

"Entirely" is an adverb that has 2 connotations. The first is "completely" or to the full extent of something, as in "I agree entirely with the decision to reject that article." The second is "solely" or to the full exclusion of something: "I am entirely responsible for the rejection of your article." "Mostly" is also an adverb and means mainly or for the greatest part of something, for example, "Mostly all case reports are now rejected by *AJNR*."

Punctuation Marks, Part 1: Colons, Semicolons, and Commas

I have grouped these because they are the condiments of written language. Because they are strong condiments, they should be used sparingly. In scientific communications, it is better to use short sentences rather than long rambling ones where ideas are separated by these marks. The "colon" is a gate that opens into a list or explanation after a statement that can stand alone. Unlike the semicolon, you nearly always know what is coming after a colon even if you have not read it. If the statement that follows a colon comprises several sentences, the first word must be capitalized. Remember: If the initial part of the statement is very brief, what follows the colon also must be capitalized.

A semicolon (;) is more difficult to use. It may be used to organize a long list as in "During my vacation I visited several cities: Rome, London, Madrid, Barcelona, Berlin, Lucerne, Vienna; and Paris, the city of lights." It may also be used to separate closely related clauses that may otherwise stand on their own. Avoid using it if you do not understand it well enough.

Commas have to be (together with periods) the most commonly used punctuation marks. They can be used to separate items in a list (generally 3 or more), to connect independent statements, to set off introductory remarks, and also in place of parentheses (think of them as "weak parentheses"). In science one does not commonly use lists of adjectives, but if you do, separate them with commas.

The fact that when you are reading something you pause does not imply you must have a comma in that place. The reasons for using (and not using) commas are vast, so it is better to use them as little as possible.

Punctuation Marks, Part 2: Parentheses, Brackets, and Dashes

If something important that you wish to emphasize does not fit neatly into your text but you feel the need to point it out, use parentheses. If the parentheses lie within a sentence, do not use a capital letter for the first word or a period after the last one (an exclamation mark may be used to further emphasize). If the statement framed by parentheses is outside of accompanying sentences, you may treat it as a freestanding sentence (start with a capital letter and finish with a period).

Using brackets is tricky. They should only be used to explain something when you are quoting some other text or person. If you are quoting and need to change the form of a word in the quotation to fit your sentence, that word should be bracketed. If you italicized or underlined a word or words, you will need to point this out in brackets [*italics mine*]. Something that is commonly bracketed in a quotation is "*sic.*" [*Sic*] generally means that the original source contains a word that is misspelled, and it follows that word. Be sure to use this sparingly; it is poor manners to call attention to the mistakes of another writer unless absolutely necessary. Last, you can use brackets inside parentheses as in "The bleed seen in the image is heterogeneous (but mostly T1 bright [Fig 1])."

Dashes are the "super commas" of language. They may also be used instead of parentheses (think of them as "super parentheses").

Use dashes when what you want to emphasize contains other punctuation marks, such as in "Certain precious metals—gold, platinum, and silver—were found to be the most common contaminants." Note that no space is to be left between the dashes and their accompanying words. If commas can do the job, avoid dashes. Dashes that frame a statement are called "em dashes" because their length is the same as the letter "M" (long dashes). Single dashes such as the ones seen between dates (February 1–3), numbers (2233–2245), and letters (items A–E) are shorter (called "en dashes" because their length is equivalent to the letter "N"). A sequence of dashes may be used when you want to omit someone's names, for example, "Professor —— did not offer a convincing argument."

As a general rule, dashes are falling from grace and it is better to avoid them.

Punctuation Marks, Part 3: Apostrophes and Quotation Marks

Apostrophes (') are generally used to create possessive forms and contractions (avoid contractions in scientific writing and speech). Apostrophes once were commonly used to create plurals, for example, "This used to happen in the 1930's." Such usage is no longer deemed necessary, for example, "This used to happen in the 1930s."

Quotation marks frame the spoken language or material that is being quoted from another source. Between them, commas, periods, and other punctuation marks may be used (this is a peculiarity of the English language not commonly seen in other languages). To complicate things even more, single quotation marks are used to frame quoted material inside other quoted material. Unless you are directly quoting from a different source, it is better to avoid them.

Some of you may think this is editorial nitpicking. Remember that how an article is written may influence the reviewers.

I have tried to give you an idea of the most commonly used (mostly erroneously) terms and punctuation marks. I have not addressed exclamation points and question marks because they are used very little in medical writing and their meaning is probably clear to most authors. When I was Senior Editor for the *American Journal of Neuroradiology* under Dr Michael Huckman, I remember using a red pencil to mark grammatic mistakes in manuscripts. Nowadays (and fortunately for us) this work is done by our editing service. Nonetheless, I hope these remarks will help some of our authors when they are writing their articles.

EDITOR'S NITPICKING # 2

Let's face it: The topic of this *Perspective* is a dry one. I promise a more entertaining one next month, but this time I think that pointing out certain problems our contributors commonly have is in order. This editorial is a continuation of my previous one, also called "Editor's Nitpicking."[1] A year after I wrote that first one, I have collected a set of words, terms, and expressions that seems to trouble authors, both English-speaking and otherwise. In addition, I've included some common Latin terms that seem to be popular with authors and are often erroneously used.

Adverbs

These words modify verbs, adjectives, and other adverbs but not nouns. Nouns, in turn, are modified by adjectives and determiners. Adverbs can be easily created by just adding "-ly" to the end of an adjective, for example, "significant" and "significantly."

Not all words ending in "-ly" are adverbs, for example, "lovely" (the root is a noun and not an adjective). Generally speaking, it is best to avoid most adverbs (and many adjectives) in scientific writing.[2] Words such as "undoubtedly,"

"unequivocally," and "substantially" overstate findings and may convey the wrong impression. The *American Journal of Neuroradiology* (*AJNR*), *Radiology,* and other journals follow the recommendations of the *AMA Manual of Style* in that "significant" and "significantly" should be used only when describing the results of statistics that reject the null hypothesis.[3]

Lay and Lie

"Lay" can be a verb or a noun. As a verb, it means to put or set down as in "Please lay your copy of *AJNR* on the table and pay attention to what I am saying." "Lay" is also the past of "lie" as in "The patient lay down before the procedure" (a sentence structure not commonly used in American English). I often see "lie" used when authors congratulate themselves, as in "The success of our technique lies in the fact that we were very careful...." "Lie" can also be a verb or noun, and when it is used as either, it generally follows the previous explanation for "lay." It is important to remember that "lie" also means to create a false statement or misrepresentation. ("The authors continuously lie about their results.") This last form has very little use in science.

There and Their

Sounds simple, no? Many of our English-as-a-second-language contributors confuse these words. "There" can be an adverb (meaning in or at that place), a pronoun (as a substitute for a name or to introduce a sentence as in "There is evidence that administering contrast is of little benefit"), or a noun (indicating place or position). "There" can also be an attribute adjective as in being fully conscious and aware of things. ("After head trauma, the patient was not fully there.")

"Their" is an adjective meaning to possess something, as in "Their MR imaging unit is superior to ours." Otherwise, when placed before a noun, it becomes an attributive adjective.

("Their rights as patients were violated by the investigators.")

Who and Whom

"Who" is a pronoun meaning what or which person or persons.

("Let's find out who developed a contrast reaction after the procedure.") Although all of us use it, strict grammarians disapprove of its use to introduce a relative clause. "Whom" is also a pronoun that appeared in the English language about the same time as "who" (12th century). "Whom" is the objective case of "who," and it is less used now than in the past.

Some historians predict that the word "whom" will eventually disappear. I still see it used often as in "Patients for whom this technique will be beneficial include those with aneurysms."

Each Other and One Another

"Each other" serves as a pronoun and is generally used when referring to 2 things that have a reciprocal relationship or action.

Conversely, "one another" is used when referring to more than 2 things. Many use "each other" and "one another" interchangeably, but strictly speaking, this is not correct.

When spoken, "each other" sounds like one word, but it is never written "each other."

Hereby and Herewith

These are adverbs, and the first means "by virtue of the present declaration, action, or document" and also "by means of this or as a result of this." "Herewith" means "along with this, together with this, or with this communication." Americans rarely use these terms, whereas our British authors employ them from time to time.

Further and Farther

"Further" is generally used when the distance it refers to cannot be exactly measured. It means "to propel or help forward, to promote, to go or extend beyond." It is related to "farther," in that it states a distance but never the exact distance, for example, "Her career will be further advanced by the publication of this important article." Conversely, "farther" is used when the distance it refers to can be quantified, as in "I can throw this ball farther than you." The confusion derives from the fact that in the past both "further" and "farther" were interchangeably used. In modern English, however, these terms have acquired different definitions and uses.

Used To and Supposed To

"Used to" should always precede a verb ("I used to live in Timbuktu"). "Used to" refers to something that happened regularly in the past but does not anymore. It is better not to use it in questions or negative statements. Sometimes "used to" can be substituted with "would to," but this sounds overly formal and is no longer commonly employed this way. "Supposed to" always carries a *d* at the end, though when spoken, it cannot always be heard (never use "suppose to"). "Supposed to" is used more often in British than American English. When "supposed to" is followed by a verb, it means "should"—for example, "I was supposed to go to the ASNR meeting, but my Chairperson did not give me permission."

Terms Expressing Time

"Today" is a commonly used adverb signifying on this day or at the present time (as in "Today, the preferred method of treating aneurysms is embolization"). When used as a noun, its meaning is the same. If used as an adjective, it means something that is characteristic of the current times. The word "now" is short but complex. It can be used as an adverb, noun, adjective, or a conjunction. It generally means at the present time or moment. Less common usages are conjunctional (meaning "in view of the fact that," as in "Now that we know gadolinium increases lesion conspicuity, it should be used in all patients"). "Nowadays" is an adverb signifying at the present time, but it is easier and more economical to simply use "today" in its place.

Numbers and Numerals

"Number" may be used as a noun or a verb. As a noun, it means the sum (or amount) of some type of unit. ("The total number of injections needed was highly variable.") This word can also be used in terms of rating as in "The number 1 neuroimaging journal is *AJNR*." It can also signify an amount as in "A large number of imaging studies were needed before reaching a correct diagnosis." When something is done in an orderly or systematic fashion, it is said to be done "by the numbers."

"Numeral" is both a noun and adjective. As a noun, it refers to the symbol for a number. Numerals can be Arabic (1, 2, 3, 4, 5, and so forth) or Roman (I, II, III, IV, V, and so forth). When used as an adjective, it relates, expresses, or consists of numbers.

"Numerically" means that there is a system or order to a series of events or numbers.

Common Latin Phrases (ibid, idem, et al, de novo, vide supra, vide infra, etc.)

"Ibid" (abbreviation for *ibidem*) is a useful term not commonly employed in scientific writing but found in other scholarly texts.[4] It means "in the same place" and is used in footnotes and bibliographies to refer to a book, chapter, article, or page cited just before. It is similar to "idem," which means something that has been previously mentioned.[4] "Et al" and "et cetera" (etc.) are used in similar fashion, but "et al" refers to a list of names, whereas "et cetera" means "and so on or more."

"De novo" means new or afresh (as in "The second aneurysm arose de novo after treatment of the first").[4] "Erratum" refers to a mistake (plural "errata") in a previous publication. "In situ" may be used to shorten the phrase "in the place that something belongs." "Per" means "through or by means of" and generally precedes another Latin term (as in "per capita").

"Prima facie" refers to evidence that is suggestive, but not conclusive, of something. "Sic" states that the preceding quoted material appears exactly that way in the source, despite any errors of spelling, grammar, usage, or facts that may be present.

[4] Be careful to use it only for important errors and not trivial ones; overuse is a nuisance. "Sine qua non" denotes something (a condition) that is an essential part of the whole.

"Status quo" is used when meaning the way things are right now or before they were upset by something or someone.

"Versus" is almost always used incorrectly (orange versus red) because it actually means "in the direction of." When we use "versus," what we really mean is "adversus." "Vide" (look or see), "supra" (above), or "vide infra" (below) are easy to understand.

I could go on and on, ad nauseam, with this editorial but I will stop here.

References

1. Castillo M. **Editor's nitpicking.** *AJNRAmJ Neuroradiol* 2010;31:1353–54. Epub 2010 Mar 4
2. Brenner RJ. **On the more insidious manifestations of bias in scientific reporting.** *J Am Coll Radiol* 2010;7:490–94
3. *AMA Manual of Style: A Guide for Authors and Editors.* http://www.amamanualofstyle.com/oso/private/content/jama/9780195176339/p175.html#jama-9780195176339-div1–215. Accessed September 22, 2010
4. Wikipedia. http://en.wikipedia. Wikipedia. http://en.wikipedia.org/wiki/Category:Latin_words_and_phrases. Accessed September 22, 2010

AUTHORSHIP AND BYLINES

From the ancient Greeks to Shakespeare, the question of authorship often arises. The issue of appropriate article authorship has always been of special interest to editors of scientific journals. In the biomedical sciences, as the complexity and funding of published studies increases, so does the length of the byline. Although a previous *American Journal of Neuroradiology* Editor-in-Chief already addressed this issue, I think it is time to revisit it.[1] From my own experience, articles can be categorized according to the number of authors as follows: fewer than 2 authors (Editorials, Commentaries, Letters), fewer than 5 authors (Case Reports and Technical Notes), 5–10 authors (retrospective full-length articles), 10–15 (prospective, often grant-funded articles), more than 15 authors (reports of task forces, white papers, etc.). Among so many authors, it is not uncommon to find individuals whose contributions are minimal and many times questionable. Who actually did enough work to be listed as an author? In other words, who can claim ownership rights in a particular intellectual property?

Academic institutions, scientific societies, and journals often have guidelines regarding authorship but, unfortunately, these are seldom respected. The International Committee of Medical Journal Editors (ICMJE) has proposed authorship guidelines.[2] The National Library of Medicine no longer limits the number of authors listed on MEDLINE (it did before MEDLINE), as long as all meet the ICMJE criteria. The Office of Research Integrity, while dealing with research misconduct, does not deal with authorship issues.[3] Although research institutions in most countries have similar offices, their involvement in regard to byline credit differs. Because works are generally the fruition of groups of individuals, these groups and their lead authors have the greatest input into the order and number of individuals listed in the byline.

The problem with this system is that many authors listed may fall into the categories of guests, ghosts, or, even worse, gifts.[4-6] "Gift authorship" is defined as "either a tribute or a ploy for recognition, with the context of reciprocal exchange or as the consequence of dependence."[7] Gift authorship is not uncommonly offered by junior individuals to senior ones through a sense of obligation. "Ghost authorship" is the opposite and refers to individuals who contributed to the work but were not listed as authors. This also commonly happens, particularly in large-scale projects in which some individuals are paid for data collection and analysis but not directly acknowledged (not an uncommon practice in industry-funded research). Not listing all of those involved is a type of plagiarism, and one study reported that it occurred in 39% of articles![8]

Several reports indicate that individual contributions are lowest in multiauthor articles, and one revealed that 26% of authors did not contribute significantly![9],[10] In the field of imaging- related journals, the *American Journal of Roentgenology* reported that undeserved authorship increased with the byline length, reaching 30% in articles with more than 6 authors.[9] On the other hand, contributors who do not want to be listed are avoiding being responsible for the integrity of an article. Research groups choosing authors or their order generally operate in egalitarian or highly hierarchic ways. Regardless of the method used to determine byline order, the implications are enormous as in most institutions order influences promotions.

Although some institutions have authorship policies, they are less rigorous than those proposed by ICMJE.[11] At any rate, editors of biomedical journals are quite serious about authorship. Some journals demand disclosure of the specific degree of author involvement, but unfortunately, most editors have little authority to enforce authorship requirements. One reason for this is a lack of consistent guidelines regarding byline listings.

There are 3 consecutive layers of byline responsibility: authors, individual offices of research integrity, and the scientific journals publishing the works. The success of each layer in monitoring appropriate authorship depends on their authority, and thus, I believe particular offices of research integrity are in the best position to monitor this issue (something they are not correctly doing).

What can be done at the author level? Communication and coordination of research at the start of a project are essential.

Of course, the ultimate byline order will be determined by the priorities and perspectives of the individuals involved. According to ICMJE, to be qualified as an author, one must meet all the following criteria: significant input into the concept and design of a study and analysis and interpretation of data, writing and revision contributions that are intellectually important, and assumed responsibility with respect to accuracy of the final contents.[2] I know of 6 journals (*American Journal of Public Health, Annals of Internal Medicine, BMJ, Lancet, Physical Therapy,* and *Radiology*) that collect information about author contributions. Some not only publish author contributions but identify those who are guarantors of the integrity of the data (crucial in multiauthor and multicenter projects).

Dissemination of data collected by these journals and widespread implementation of contributorship systems may lead to greater responsibility.[3] One study estimated at least one third of journals do not adhere to these guidelines, whereas another found that though 64% of authors met the guidelines, they were not familiar with them.[12]

When contributors do not meet criteria to be credited as authors, it is common to list them at the end of articles in acknowledgments. It has been proposed that acknowledgments be reserved for individuals with limited or purely technical contributions.[13] This leaves the question of how to recognize contributors who fall in the middle, such as those providing patient care. Weighing of author contributions is generally a purely qualitative assessment. The family practice and biostatistics disciplines experimented with qualitative weighing of contributorship with little success.[14],[15] One study looked not only at the number of authors but at their academic ranks and found that most authors were either professors or residents.[16]

To give credit to all those involved, dichotomous and trichotomous author categorizations have been suggested.[17] Using this type of system, the concept of one author making a unique contribution would cease.[6] Additionally, others have considered making publications anonymous or listing authors alphabetically, without success.[17] I have been asked by several contributors about the possibility of listing 2 individuals as first authors, and a dichotomous system would allow us to list a "first" author in the category of the work in which each contributed most. After pondering these systems, I have decided, for the time being, to keep our traditional, simple, 1-level author listing. In order for dichotomous or trichotomous listings to be meaningful, promotion committees and funding agencies would have to recognize these first. One last system has been suggested at the author level: weighing of contributions by a "third" disinterested party. This method may fall into the responsibilities of specific offices of research conduct.

Last century, deconstructionists attempted to break down texts to observe who coveted power and how.[17] In science, we all have witnessed power struggles when it comes to credit for publications. The responsibility of journals for bylines is difficult to assess and impose. Confronting author credit and responsibility is a daily predicament for editors. I have been pleased by the fact that when asked about long bylines, our contributors have always responded responsibly by shifting an excessive number of individuals into acknowledgments or by clearly justifying their degrees of involvement. Author responsibility should be shared by authors, their institutions, and journal editors. Our credibility as researchers depends on this type of responsibility and avoiding abusing it.

References

1. Quencer RM. **Creeping authorship: where do we draw the line?** *AJNR Am J Neuroradiol* 1998;19:589
2. International Committee of Medical Journal Editors. **Uniform requirements for manuscripts submitted to biomedical journals: writing and editing for biomedical publication.** Available at: www.icmje.org. Accessed March 17, 2009

3. Council of Science Editors. **Publications**. Available at: www.council-science editors.org/publications/v23n4p111–119.pdf. Accessed March 17, 2009
4. Rennie D, Flanagin A. **Authorship! Authorship! Guests, ghosts, grafters, and the two-sided coin.** *JAMA* 1994;271:469–71
5. Flanagin A, Fontanarosa PB, Phillips SG, et al. **Prevalence of articles with honorary authors and ghost authors in peer-reviewed medical journals.** *JAMA* 1998;280:222–24
6. Rennie D, Yank V, Emanuel L. **When authorship fails: a proposal to make contributors accountable**. *JAMA* 1997;278:579–85
7. Council of Science Editors. **Authorship task force**. Available at: www.councilscienceeditors.org/services/authorship.cfm. Accessed March 17, 2009
8. Mowatt G, Shirran L, Grimshaw JM, et al. **Prevalence of honorary and ghost authorship in Cochrane reviews.** *JAMA* 2002;287:2769–71
9. Slone RM. **Coauthors' contributions to major papers published in the *AJR*: frequency of undeserved authorship.** *AJR Am J Roentgenol* 1996;167:571–79
10. Shapiro DW, Wenger NS, Shapiro NF. **The contributions of authors to multiauthored research papers.** *JAMA* 1994;271:438–42
11. Jones AH. **Is the system really broken?** *Lancet* 1998;352:894–95
12. Hoen WP, Walvoort HC, Overbeke AJ. **What are the factors determining authorship and the order of authors' names?Astudy among authors of the Nederlands *Tijdschrift voor Genesskunde* (Dutch *Journal of Medicine*).** *JAMA* 1998;280:217–18
13. Fotion N, Conrad CC. **Authorship and other credits.** *Ann Intern Med* 1984;100:592–94
14. Ahmed SM, Maurana CA, Engle JA, et al.**Amethod for assigning authorship in multi-authored publications.** *Fam Med* 1997;29:42–44
15. Parker RA, Berman NG. **Criteria for authorship for statisticians in medical papers.** *Stat Med* 1998;17:2289–99

16. Drenth JPH. **Multiple authorship: the contributions of senior authors.** *JAMA* 1998;280:219–21
17. Council of Science Editors. **CSE task force on authorship.** Available at: http:// www.councilscienceeditors.org/services/atf_whitepaper.cfm. Accessed March 17, 2009

AUTHORS' NAMES

We at the *American Journal of Neuroradiology* (*AJNR*) have the responsibility of correctly publishing the names of our authors. Apart from being a courtesy to them, adequate and consistent name presentation (particularly from authors who repeatedly contribute) leads to their recognition, easier accessibility to their articles, and increased citations. Lack of uniformity is generally not an issue with English names, but with *AJNR* becoming more and more international, understanding and correctly displaying the names of our authors can be complicated. I urge our authors to follow some simple rules when submitting their articles to *AJNR* to ensure that their names are correctly mentioned. *AJNR* generally publishes only initials for first and middle ("given") names and spells out last ("family") name(s). Ordering of names is generally different from country to country and follows rules that are generally historical and cultural in nature. Let me briefly review the structure of names from the zones of the world where most of our authors originate. To avoid confusion, I will use the terms family name instead of last name or surname and given name instead of first name or middle name.

Anglophone names. These generally comprise 1 or 2 given names (first and middle) followed by the paternal family name (e.g., Robert Francis Kennedy). In some cases, the maternal family name may be used as a middle name, but this is mostly a personal preference. On occasion, individuals may have only a first but no middle name (e.g., Edward Jenner).

Chinese names. The Chinese use different systems that allow them transliteration of their names to English, such as pinyin and the Wade-Giles system.[1] The first is mostly used in mainland China and the second in Hong Kong. Their family name precedes their given name(s). Family names are generally monosyllabic (e.g., Wang) but occasionally contain 2 syllables (i.e., Ou-Yang), whereas given names may have 1 syllable (e.g., Song) or 2 syllables (e.g., Zhi Yong). Given names may be spelled separately or hyphenated. Most Chinese authors submit their names in English format, but when they send it to us in its original order, confusion may arise.

Korean names. In Korea, the family name comes first and is followed by a duo syllabic given name (e.g., Kim Daejung).[2]

Family names are generally monosyllabic but occasionally they contain 2 syllables. Transliteration to English is generally done by use of the McCune-Resichauer system or the government- created Revised Romanization of Korean system. When most Koreans write their names in English, they follow Western order (this is true in most medical journals).

Japanese and Vietnamese names. In Japan and Vietnam, the family name also comes first and is followed by a given name.[3] Many Japanese family names take root in features of rural landscape and are extremely varied. Middle given names are not used in Japan, but when used, they become one with the first given name because a space is not a permitted character.

Middle given names are, however, used in Vietnam. Although given and family names are easily recognizable in Japanese, they may coincide and become indistinguishable when romanized. Most Japanese and Vietnamese academics whose work is published in English journals follow the Western order when listing their names.

Spanish and Portuguese names. These names follow a more complicated structure. Spanish persons may have 1 or 2 given names followed by 2 family names. Family names are organized as follows: paternal first and maternal second. Some of us have "Americanized" our names (I legally changed mine from *Mauricio Castillo Gonzalez* to only *Mauricio Castillo*) to avoid confusion. For Spanish-speaking authors, I have the following suggestion: use only your paternal family name or hyphenate both family names (look at the names of the authors listed below in reference 5).[4,5] If I had retained my original name, it is possible that I would be mentioned as *M.C. Gonzalez* in the literature. A married woman may adopt her husband's paternal family name and drop her maternal family name but retain her paternal one. This is evident to readers because the newly acquired family name will be separated from the previous one by the preposition *de* ("of"), as in *Maria Victoria Mendoza de Perez* (in *AJNR*, her name would be listed as *M.V.M. de Perez*). This tends to create a problem when *de Perez* is dropped, and she becomes, again, *M.V. Mendoza Rivas*. In Portuguese, family names are also compound, but the maternal name comes first. As a tradition, when a woman marries, she drops the maternal name and keeps the paternal one while adding her husband's paternal last name. When Portuguese names are indexed, their last family name is used.

Indian names. As a tradition, Indians did not have the concept of using family names, but during British occupation they integrated them into family life.[6] A traditional Indian name comprises the name of the native (ancestral) village, father's name, given name, and sometimes the caste title (i.e., *Rasipuram Krishnaswami Ayyar Narayanan*). Married women may use their husband's name as their family name. This is particularly true in the southern regions of India. For example, *C.V. Raman* is a famous physicist, but *Raman* is his given and not family name. We urge our Indian contributors to list their family name last so it can be correctly quoted. Send us your name exactly as you wish to see it in *AJNR*.

Arabic names. Traditional Arabic names consist of 5 parts (but may have more) as follows: given name (in males it may be preceded by one of the attributes of Allah), honorific name (generally does not appear in print), paternal family name (starting with *bin* or *ibn* meaning "son of" or "daughter of," respectively), a religious or descriptive epithet, a last name that may be a true family name or a short phrase that stands for occupation, geographic location, or tribe.[7] This complex traditional practice is declining, particularly in countries with Western influences such as Lebanon and some African countries.

Women do not take their husband's family name when they marry. Christian Arabs may use a combination of names derived from the Bible.

I have not specifically mentioned Western European names because most French, German, Italian, and other names comprise a simple or compound given name and a patronymic family name. English is not the most commonly used language in the world, but it is certainly the de facto language for science. More than 90% of all scientific and medical journals in the world are published completely or partly in English.

We at *AJNR* do our best to publish author's names correctly, but because of an increasing number of articles originating from outside of the United States, we urge authors to take responsibility and to let us know exactly how they want their name to appear. In this editorial, I have attempted to give our readers an idea about the complexity involved.

REFERENCES

1. Sun XL, Zhou J. **English versions of Chinese author's names in biomedical journals: observations and recommendations.** *Science Editor* 2002;25:3–4
2. Han S. **Formats of Korean author's names.** *Science Editor* 2005;28:189–90
3. **Japanese name.** From Wikipedia, the free encyclopedia. Available at: http:// en.wikipedia.org/wiki/Japanese_name. Accessed September 12, 2008.
4. Black B. **Indexing the names of authors from Spanish- and Portuguese-speaking countries.** *Science Editor* 2003;26:118–21
5. Ruiz-Perez R, Delgado Lopez-Cozar E, Jimenez-Contreras E. **Spanish personal name variations in national and international biomedical databases: implications for information retrieval and bibliometric studies.** *J Med Libr Assoc* 2002;90:411–30
6. Kidambi M. **Indian names: a guide for science editors.** *Science Editor* 2208;31:120–21
7. Notzon B, Nesom G. **The Arabic naming system.** *Science Editor* 2005;28:20–21

FUTURISM AND SCIENTIFIC NETWORKING

In a plenary session at the last meeting of the Council of Science Editors, Mr. Blake Godkin made the following comment:

> "In a time of accelerating change—when the future of any nation will be based on how well it creates and manages new ideas—we can no longer make current decisions

primarily on the basis of experience. This is why it is imperative to become comfortable with the process of thinking like a futurist."

[1] Mr. Godkin, I might add, works for the SHW Group, a Texas-based firm of architects that designed the Dubai extension of the University of Michigan and parts of the Texas Tech University campus; they specialize in designing education related buildings. Architecture has historically been involved with futurism, as has literature. The prediction of the future also plays a critical role in medicine and in scientific publishing.

The term *futurism* broadly means to speculate about the future. Initially referring to a religious movement, futurism became popular in the early 20th century, and the term was appropriated by architects, painters, and other people associated with the avant-garde movement. In literature, the futurism movement started in Italy. From people, futurism extended to other areas and was later adopted by consulting companies such as the RAND Corporation and others that mainly deal with prediction of military conflicts and their consequences.

Today's futurism (or more correctly, *neofuturism*) is represented by groups of people who study and attempt to predict global trends mostly as they relate to businesses and man-made or natural disasters. The formal study of the future includes foresight, strategy, and prospective awareness. Successful futurologists use trend, precursor, and scenario analyses in their prospecting.

In my opinion, scientific journal editors have to be imaginative and be driven futurologists. Systematic collections of data indicating a trend are commonly used in medicine to predict the future of 1 or more events. Unfortunately, predicting long-term trends is fraught with considerable problems.

Precursor analysis takes into account that phenomena go through changes and, thus, we should be able to anticipate them. Creating diverse scenarios enables us to anticipate phenomena.

The World Future Society[2] emphasizes all of these mechanisms and believes that continued improvement in how they are used will lead to better understanding and planning for the future. Of course, to achieve these goals, creativity is also of the essence.

In the previously mentioned lecture, Mr. Godkin also told the audience what he considers to be the tools of the futurist: brainstorming, trend analysis, and trend scanning (these are also the sequential stages needed to make assumptions regarding the future). As futurologists, at the brainstorming stage we need to accept all ideas, no matter how wild they may initially seem. Unrelated thoughts may give origin to new and valuable ones. The term *brainstorming* was probably coined by Alex Osborn, a mid-20th century advertising executive who characterized it as encompassing the following: no criticism of ideas, large quantity of ideas, building on of other ideas, and encouragement of wild and exaggerated ideas.[3] Mr. Osborn said, "It is far easier to tame a wild idea than to invigorate a dull one." Many of the options available on *AJNR's* Website are the fruition of successful brainstorming. Prospective trend analysis spots a pattern and takes advantage of it. Trend scanning is currently easier than in the past because of the use of search engines. By analyzing key terms entered into search engines, we can understand what our readers and Website users want.

Starting our blog site (www.ajnrblog.org) and the Case of the Week are examples of the results of trend scanning and adapting to what the public needs and wants.

What are the advantages of having a blogsite? Most internet users are young (between 18 and 44 years old), and trend research indicates that e-mail is used much more by older vs younger people.[4] E-mail is the most popular internet activity among older persons, and nearly 75% of persons 64 years or older use it regularly. In the last 4 years, the usage of e-mail among younger people has decreased by more than 15%. Persons aged 18 to 32 years (Generation Y) use telephone texts, blogs, and other social networking activities as their main forms of communication and not e-mail. As younger generations become incorporated into the American Society of Neuroradiology and *AJNR* readership, blogging will become an important activity; it is not surprising that older generations rapidly learn how to blog and also enjoy it.

A blogsite is not exactly the same as an internet forum or message board, though, to me, the differences are minimal. An internet forum and a message board are discussion sites perhaps analogous to the bulletin board of yore. A blogsite's contents are more structured than those of forums and boards.

These networking sites may create a paradox that is important in science and is called *participatory journalism*. It implies that news is no longer provided by a group of trained journalists and writers but by the public in general. The risk for this trend in science is that unsupported ideas and concepts become slowly incorporated and, with the passing of time, are assumed to be truths. Anonymous networking sites encourage this type of behavior, which is why our blogsite requires all contributors to clearly identify themselves. Although most blogsites serve as places of social networking, ours should serve as a place of scientific networking.

Being highly specialized, as we neuroradiologists are, limits one's exposure to assessing the future as a whole. Thus, we must seek what parts of futurology we need to practice. For ourselves on the editorial board and for our readers, we must be imaginative, agenda-driven, consensus-driven, critical, alternative, predictive, and validating futurists. Epistemologists tell us that knowledge lies at the crossroads of truths and beliefs.

The future of knowledge will, in part, depend on constant interactions such as the ones possible on our blogsite.

References

1. Godkin BS, Pagel WJ. **Plenary address: thinking like a futurist.** *Science Editor* 2009;32:4
2. World Future Society. Available at: www.wfs.org. Accessed February 2, 2009
3. History and use of brainstorming. Available at: http://www.brain storming.co.uk/tutorials/historyofbrainstorming.html. Accessed February 6, 2009
4. Available at: http://www.readwriteweb.com/archives/whos_online_and_what_ are_they_doing_there.php. Accessed February 17, 2009

www.ingramcontent.com/pod-product-compliance
Lightning Source LLC
Chambersburg PA
CBHW072301200526
45168CB00014B/78